Chemical misconceptions

– prevention, diagnosis and cure

Volume I
Theoretical background

Written by Keith Taber
RSC School Teacher Fellow 2000–2001

ROYAL SOCIETY OF CHEMISTRY

Chemical misconceptions
– prevention, diagnosis and cure
Volume 1
Theoretical background

Written by Keith Taber

Edited by Colin Osborne and Maria Pack

Designed by Imogen Bertin

Published and distributed by Royal Society of Chemistry

Printed by Royal Society of Chemistry

For further information on other educational activities undertaken by the Royal Society of Chemistry write to:

Education Department
Royal Society of Chemistry
Burlington house
Piccadilly
London W1J 0BA

Information on other Royal Society of Chemistry activities can be found on its websites:
http://www.rsc.org
http://www.chemsoc.org
http://www.chemsoc.org/LearnNet contains resources for teachers and students from around the world.

ISBN 0–85404–386–1

There is a companion volume to this publication, *Chemical misconceptions – prevention, diagnosis and cure*: *Volume 2 Classroom resources* ISBN 0–85404–381–0
The combined ISBN number for both volumes is: 0–85404–390–X

British Library Cataloguing in Publication Data.

A catalogue for this book is available from the British Library.

RS•C

Foreword

Chemistry is a conceptual subject and, in order to explain many of these concepts, models are used to describe and explain the microscopic world and relate it to the microscopic properties of matter.

As students progress in chemistry the models they use change and many contradict their everyday experiences and use of language. This resource is designed to explain some of the misconceptions students have.

It is hoped that teachers will see the benefit of a constructivist approach and find that the companion volume of student material leads to a good theoretical underpinning of the fundamentals of chemistry.

Professor Steven Ley CChem FRSC FRS
President, The Royal Society of Chemistry

RS•C

Contents

RS•C

RS•C

Acknowledgements

I would like to acknowledge the cooperation of a large number of people and organisations who helped with this project in various ways. The following colleagues and institutions trialled or commented on draft materials, or sent useful suggestions or information. I hope I have remembered everyone, but with such a collective effort I trust any omissions with be forgiven.

Glen Aikenhead – University of Saskatchewan, Canada
Jon Angell – Downend School
Stuart Barker – Park College
Michael Barton
Richard Biddle – Carmel Technology College
Andrew Biggs – Shrewsbury School
Peter Biggs – Hinchingbrooke School
Vanessa Bird – Worth School
Marc Bloch – Heinemann Educational Publishers
John Bloor – Universities of Virginia and Tennessee, USA
Martin Bluemel – Taunton School
Anne Brearley – United World College of the Adriatic, Italy
Peter Buck – University of Education, Heidelberg, Germany
Mike Clugston – Tonbridge School
David Cooper – Sutton Valence School
Di Cook – Wispers School
Andrew Davies – St Boniface's College, Plymouth
Gaynor Davies – Ounsdale High School
Philip Dobson – Berkhamsted Collegiate School
Michael Dorra - University of Education, Heidelberg, Germany
Keith Fleming – University of New England, Australia
Duncan Fortune – Glasgow Caledonian University
Mike Foster – St Margaret's School, Exeter
Mark Gale – South Dartmor Community College
John Gilbert – University of Reading
Grant Gill – Kemnal Technology College
Alistair Gittner – King's School, Peterborough
Alan Goodwin – Manchester Metropolitan University
Richard Grime – The King's School, Macclesfield
Anthony Hardwicke – Monmouth School
Claus Hibling – Institute for the Teaching of Chemistry, Münster, Germany
Geeske van Hoeve-Brouwer – Eindhoven University of Technology, The Netherlands
Andrew Hunt – Nuffield Curriculum Projects
Anne Hurworth – Park Lane College, Leeds
June Jelly – Menzieshill High School
Bill Johnson – International School of Geneva
Philip Johnson – University of Durham
John Kerr – Winchester College
Ros Key-Pugh – Royal High School
K. G. King – Perse School for Girls
Sarah Knight – James Allen's Girls' School
Gill Kuzniar – Netherhall School
Michael Laing – University of Natal
Helen Langslow – Stamford High School
Mark Leach

RS•C

Christine Lewis – Keswick School
Steve Lewis – Shrewsbury Sixth Form College
John Luton – Varndean College, Brighton
Robin Millar – University of York
Roger Mitchell
Raphael Mordi – Newcastle College
Matthew Morrison – Strathcona Baptist Girls' Grammar School, Australia
Brian Murphy – United Arab Emirates University
Jim Murphy – St. Boniface's College, Plymouth.
Paul Murphy – Loretto School
Vincent Murphy – Copenhagen International School
David Neill – Canford School, Wimborne
Igor Novak – National University of Singapore
Jerry O'Brien – Slough Grammar
Philip O'Connor – Oakwood Park Grammar School
John Oversby – University of Reading
Andrew Page – Braintree College
Lambros Papalambros – Livadia High School, Greece
Nicole Pearce – Headlands School
M. Arminda Pedrosa – University of Coimbra, Portugal
Margaret Price – Balfron High School
K. C. Pun – Tsang Shiu Tim Secondary School, Hong Kong
Alan Quinn
David Reynolds – Long Road Sixth Form College, Cambridge
Graham Riley – King Edward VI Grammar School, Chelmsford
Keith Ross – Cheltenham and Gloucester College of Higher Education
Wendy Rudge – Canterbury College
Charly Ryan – King Alfred's College, Winchester
Eric Scerri – University College of Los Angeles
Hans-Jürgen Schmidt – University of Dortmund, Germany
Colin Smith
Mrs. S. Smith – Hazelwick School
Marinella Spezziga
Lesley Stanbury – St. Albans School
Julia de Ste Croix – East Devon College of Further Education
Ann-Marie Stott – St Paul's Catholic College, Haywards Heath
Ben Styles – University of Sussex
Joanne Sumner – Maricourt High School
Mohammed Taj – Deacons School
Daniel Tan – Nanyang Technological University, Singapore
Zoë Thorn – Saffron Walden County High School
Tony Tooth – King's School, Ely
Georgios Tsaparlis – University of Ioannina, Greece
Zoltán Tóth – University of Debrecen, Hungary
Nicos Valanides – University of Cyprus
Christina Valanidou – University of Cyprus
David Waistnidge – King Edward VI College, Totnes
Andrew Wallace – Durham School
Paul Warren – Linton Village College
Simon Warburton – Ferrers School
Mike Watts – University of Surrey Roehampton
Jonathan Wilkinson – Bingley Grammar School
Bob Wright – Charterhouse School

RS•C

I would also like to thank all the correspondents on the learning-science-concepts e-mail discussion list (**http://uk.groups.yahoo.com/group/learning-science-concepts**, accessed September 2005) for some stimulating and provocative exchanges about some of the issues discussed in this publication.

Most of the classroom materials presented in this resource pack were written by the author, although informed by published or other research. The probe on Mass and dissolving was based on a question written by Dr. Vanessa Barker of the Institute of Education, University of London. The probes on Elements, compounds and mixtures were based on a question set as part of a National Survey undertaken by the Assessment of Performance Unit of the (then) Department of Education and Science in the U.K. The other classroom materials are original, but informed by a variety of existing research which is cited in the text.

Above all, thanks are due to the Royal Society of Chemistry for funding the project and awarding the Teacher Fellowship; Homerton College (and in particular, the Principal, Dr Kate Pretty) and the Faculty of Education, University of Cambridge for releasing me from teaching duties; The University of London Insitute of Education for awarding a Visiting Fellowship, and the Science & Technology Group there for providing a welcome, comradeship and logistical support; Dr Colin Osborne (RSC Education Manager, Schools and Colleges) and Dr Maria Pack (RSC Assistant Education Manager, Schools and Colleges) for support and for trusting my instincts; members of the RSC Committee for Schools and Colleges for commenting on drafts, and Philippa Taber for some clerical assistance with data analysis, but mostly for generous tolerance of my obsessions.

Keith Taber
August 2001

RS•C

How to use this resource

Electronic versions of the worksheets in the companion volume are available at **www.chemsoc.org/networks/learnnet/miscon2.htm** which may be downloaded and customised.

Quotes from journals or books are indicated using speech marks.

Quotes of speech are shown in italics and speech marks.

[*sic*] – this word is used to indicate an apparent mispelling or doubtful word or phrase in the source being quoted.

Each chapter begins with a summary.

The chapters of this volume, Theoretical background, are linked to the probes published in the companion volume, Classroom resources, as follows:

Theory chapters	Probes
1	Elements, compounds and mixtures
2	Changes in chemistry; Definitions in chemistry; Acid strength; Chemical comparisons; Elements, compounds and mixtures
3	Revising acids; Revising the Periodic Table; Scaffolding explanations
4	Learning impediment diary
5	Revising acids; Scaffolding explanations; Precipitation, Elements, compounds and mixtures
6	Elements, compounds and mixtures; Mass and dissolving; Definitions in chemistry; Revising the Periodic Table; Iron - a metal; Precipitation; Chemical stability; Stability and reactivity; Changes in chemistry
7	The melting temperature of carbon, An analogy for the atom; Ionisation energy; Chemical comparisons
8	Iron - a metal; Ionic bonding; Precipitation; Spot the bonding; Interactions
9	Word equations; Types of chemical reaction; Precipitation; Hydrogen fluoride; Reaction mechanisms
10	Hydrogen fluoride

All the probes mentioned in this publication are shown in bold and can be found in the companion volume *Chemical misconceptions, prevention, diagnosis and cure – Volume 2, Classroom resources.*

RS•C

Teacher's feedback form

**Chemical misconceptions – prevention, diagnosis and cure.
Volume 1 Theoretical background**

The Royal Society of Chemistry welcomes feedback on how useful teachers find this publication.

If you have any comments on the book, or on specific resources or suggestions included in it, please make a copy of this page and send your comments to

Dr. Keith Taber
c/o Education Department
The Royal Society of Chemistry
Burlington House
Piccadilly
London W1J 0BA

Please include an email address if you would like an acknowledgement of, or response to, your comments.

Teacher's name:

_____

Institution name:

_____

Email address (optional):

_____

Your comments:

_____

_____

_____

_____

_____

_____

_____

_____

_____

_____

_____

_____

_____

_____

Thank you for taking the time to let us know your views on these materials.

RS•C

1. Alternative conceptions in chemistry teaching

This chapter provides a brief introduction to the topic of learners' ideas in science, and in particular to the types of alternative conceptions that have been uncovered among chemistry students in schools and colleges.

Students' alternative ideas

Water turns blue when copper sulfate is added:

'because the copper sulfate has a chemical inside it that turned the water blue'
a student in class of 11–12 year olds

A compound is:

'a substance that contains 2 or more of the same kind of atom'
a student in a class of 13–14 year olds

Hydrogen reacts with fluorine:

'because fluorine wants to gain an electron and hydrogen wants to lose one'
a student on post-16 course

The above comments were made by students responding to probes that are included in this publication. These comments, and many others like them reported in the research literature, show that students often develop alternative ideas about the science they are taught in school.

What learners don't know about science

Nobody is very surprised that when students are asked questions about science topics which they are meant to have studied, they often give 'wrong' answers. If matters were otherwise then presumably students would all score 100% on tests, and there would be no need for science teachers to be highly skilled classroom practitioners (and little need for the tests either). This much is true in other subjects as well as science, and may be explained in a number of ways. Assuming students were actually in the lesson when material was taught then they may not have been paying attention; or they may have not understood what the teacher said; or they may have just forgotten. As we are all sometimes guilty of not paying attention; and there are times when we do not understand what is said to us; and as we all sometimes forget things we wanted to remember, this may seem to be explanation enough.

Teachers are usually aware when students are not paying attention, and respond accordingly. Teachers also have ways of finding out when learners do not understand. In an ideal world we create the type of supportive learning environment where students are keen to learn, take responsibility for asking when unsure, and are confident to speak up without feeling self-conscious or in danger of ridicule.

Even when our classrooms and teaching laboratories do not match this ideal, teachers learn to use questioning techniques to check learners' understanding.[1] As effective learning requires regular reinforcement and review, forgetting is a more difficult problem for the teacher to tackle, particularly where there is much material to cover and classes are not seen frequently. However, even in this area, teachers can take every opportunity to bring in previous work and relate this to new topics, and can encourage effective study skills in their students.

No doubt such professional techniques are very useful, but – even with the improvements they may bring – we do not expect all the class to get near perfect marks on the end of topic tests.

RS•C

What learners think they know about science

What may be more surprising than students' failures after teaching are their responses when they are asked questions about science topics before they have covered the work. Clearly students are less likely to produce acceptable scientific responses before teaching than afterwards. (It would be very worrying if this was not the case.) What is less obvious is how often the learner is able to produce some kind of answer to scientific questions before they have (formally) learnt anything about the topic.

Of course sometimes youngsters may just feel they ought to think of an answer because of the social pressure of being asked by an adult.[2] The influential Swiss psychologist Jean Piaget described how young children would often 'romance' up an answer when they were asked something that they did know about.[3] However, research evidence suggests that many of the answers that students produce cannot be explained in that way and there is a great deal of research evidence available to consider.

Over the past twenty years there has been a vast research effort to ask students all sorts of questions about many different science topics - before, during and after teaching. There are now thousands of papers in research journals and conference proceedings presenting the results of this research, and a range of books discussing the findings.[4] In view of the great diversity of this work (undertaken in the UK, New Zealand, Australia and many other countries; with learners of various ages from young children to graduates; and in aspects of biology, physics and chemistry) it is not surprising that the 'experts' do not all agree on all the details of what this research tells us. However, there is a general consensus on many important points, and it is certainly agreed that learners hold a wide range of ideas about many scientific topics – ideas that often contradict the science they will meet in school and college.[5]

Students as scientists?

It is certainly very clear that teachers can not safely assume that their students will come to classes without any preconceived ideas about a topic, giving the teacher a 'blank slate' on which to impress scientific knowledge.

The late Rosalind Driver (who was very influential in undertaking and encouraging research into learners' ideas) described 'the pupil as scientist', and explained that from a young age children behave like amateur scientists, finding patterns in the world and forming conjectures to explain these patterns.[6]

Of course, the youngsters are not professional scientists, and so their thinking does not always match scientific standards.[7] The important point was that by the time a student comes to secondary science he or she will have built up a great many of their own explanations about the way the world works, and many of these will be at odds with the scientific view. (The student will also have studied science in primary school, but may well – despite having being given satisfactory explanations – have misinterpreted that teaching in the terms of their prior conceptions).

One way in which Driver found that students were rather poor scientists was in the way they treated data. Driver found that students were often unable to see that the results of an experiment should have refuted their ideas about what was going on. Indeed, she found that often students would 'see' and record what their preconceptions told them to expect to see – and so their recorded results matched their expectations rather than what they were meant to observe. (If one wished to be cynical one might suggest that Driver got her slogan wrong – and that her book should have been called 'the pupil as politician', 'the pupil as Freudian analyst' or 'the pupil as Marxist historian'.)

The important point that Driver recognised was that this failure to record results accurately was not due to laziness, or stubbornness or being deliberately awkward. Students were not being arrogant – just human. Driver's observations reflected something very important about the human perceptive system: we often see what we expect to see rather than what is in front of our eyes. From a physiological point of view we do not see with our eyes, but with our brains, and the signals from the

RS•C

eyes are only part of the information being used to make sense of the world. This can be seen by referring to one of the books presenting optical illusions,[8] which show how our brains attempt to interpret what we see in terms of the patterns we expect.

Professional scientists receive training in applying practical and analytical techniques to help them learn to give precedence to the data, and to be (less) biased by their expectations in experimental work. Students in school are only just starting out on their training in scientific method.

The nature of learners' ideas

The research into learners' ideas about scientific topics suggests that students' informal ideas vary across a number of dimensions. Some of the ideas reported appear to be quite specific, whilst others appear to be more general. Sometimes learners' ideas are quite labile, and are readily changed, but others may be very stable, and quite tenacious in the face of instruction.[9] Some of the ideas reported seem to be fairly isolated and not particularly well related to the students' other ideas, whereas some ideas seem to be embedded in complex, coherent, logically related structures.

Certain researchers who have worked in the field have suggested that because some of the alternative ways of explaining science revealed in research have seemed to be loosely associated clusters of ideas or logically incoherent, then learners' ideas are always like this.[10, 11] In my own research I have found that learners' thinking can be very variable. Students, like any of us, can have strongly held beliefs, as well as vague notions, and relatively isolated ideas as well as logically developed frameworks of conceptions.

The advice I would give to teachers can be summarised in three points:

1. In any class, for any science topic, students are likely to hold a wide range of alternative ideas about the topic;

2. not all of these ideas will be highly significant in terms of impeding the intended learning; but

3. some of them will.

Therefore, teachers need to take learners' ideas seriously.

The significance of learners' ideas

Some of the ideas that students will bring to class will just appear to 'evaporate' away once the scientific perspective is presented. However, this is certainly not always the case. Indeed there are a number of possible outcomes when a student (holding one set of ideas) is taught science which is inconsistent with their prior conceptions;[12]

1. Sometimes the new learning does seem to successfully supplant the old ideas without too many problems.

2. Sometimes the learner treats the new ideas as if they are unrelated to their previous thinking. The science that has been learnt in school seems to be 'stored' separately from the existing ideas. Sometimes when this happens the student may use one set of ideas to answer formal science questions, but a different set of ideas in everyday situations.

This research suggests that when the same scientific principle is tested in an abstract 'scientific' context, it will be answered differently to an equivalent question testing the same principle, but set in a novel everyday context.[13, 14, 15, 16, 17] This may have consequences for examiners looking to set questions in novel everyday contexts,

'[A] question on differences between the element iron and its compounds... set in the context of breakfast cereal content was... [often] answered using everyday understanding rather than scientific knowledge, for example stating that iron is inedible, whereas iron compounds are edible.'[18]

RS•C

Some science educators feel it is satisfactory for learners to acquire separate versions of scientific ideas which they apply in academic and everyday (or 'life-world') contexts, and some students seem to be able to successfully retain workable versions of the ideas in both of these 'domains'.[19, 20]

3. However, it is often found that when the student stores the new ideas separately they are soon forgotten. Although tests shortly after the teacher's presentation may be encouraging, it is sometimes found that re-testing weeks and months later leads to the student returning to their original way of thinking about the topic.[21] This is disappointing as the learning that occurs appears to have been rather superficial (and does not survive to end of course examinations, let alone through later life). Indeed it may be found that experiments that were presented to demonstrate the scientific ideas are now recalled as having different results – results which support the students' original way of thinking about the topic. Retrieval of information from memory is known to be a reconstructive process, where much of a 'memory' is often inferred from specific items of recall.[22,23]

4. Sometimes little or no learning takes place because the learner is unable to make sense of the teacher's presentation in terms of their existing ideas. In this case the learner's ideas may seem unchanged by the lesson.

5. Sometimes the learner is able to make sense of the teacher's presentation in terms of their own alternative way of thinking about the topic. This may result in the student learning new material, but not in the way intended. The learner unintentionally distorts the teacher's words to fit into the existing framework. Often, when this happens, neither the teacher nor the student are aware that the student is reinterpreting the material in this way – at least not until the new learning is elicited in a test.

This may mean that there is no *fundamental* change in the way the learner understands the topic, despite new learning having taken place (as the new ideas are all made to fit with the existing understanding). However it is also possible for the process of 'making sense of the teacher' to lead to the student's ideas starting to change. The result may be a hybrid understanding of a topic somewhere between what the student started out with, and what the teacher intended.

As we all know, teaching science is a challenging and complex affair.

Describing learners' ideas

Up to this point I have deliberately been a little vague in the way I have described learners' ideas about science topics. This is because, unfortunately, experts do not agree on what terms should be used.[24]

If you read some of the journal articles and books in the Notes and references section you will find references to a whole range of terms. Students' ideas may be described as intuitive, informal, misconceived, alternative, preconceived, prior, folk, life-world, etc; they may be ideas, concepts, conceptions, frameworks and so on.

The reason why so many terms are used is (in my view) because learners' ideas are so varied. Some aspects of thinking may well reflect the structure of the human cognitive apparatus (*ie* the way nerve cells in the brain work together), and could be considered 'intuitive'. Other ideas are picked up from the social milieu – the playground, television, listening to parents and older siblings etc – and may be described as 'informal'. If a student misinterprets what a teacher has said we might call their idea a 'misconception'.

Of course, many ideas that learners have can not be so easily classified. Brain structure; early experience of the world; the quirks of language; things heard, seen and read out of school; and classroom experiences may all play a part in building up new ideas. All new learning is interpreted through existing ideas, so few notions that people have can be said to derive from just one source.

RS•C

I will refer to learners' ideas that do not match science as being 'alternative', that just means they are different, without needing to consider how they arise. (Later it will be suggested that it is useful to distinguish those alternative conceptions which seem to derive partly from the way we teach topics, to those which students seem to acquire regardless of how we teach – see Chapter 4.)

Two terms that are commonly used are 'alternative conceptions' and 'alternative frameworks'. I tend to use these terms to have slightly different meanings so that;

- alternative conception - refers to a single idea; and

- alternative framework - refers to a complex or structure of related ideas.

There are some examples given below. However, you should bear in mind that it is not always obvious whether an elicited conception is actually part of a more complex framework, and in some articles and books you will find these terms are used interchangeably, as if they are synonymous. It is less important which terms we use to describe students' ideas, than to (a) recognise that students have alternative conceptions that may interfere with learning, and (b) know how to diagnose and try and respond to them. That is what this publication is about.

Some examples of alternative conceptions

Alternative conceptions are found in all areas of science. For example, in physics, it is found that something like 85% of secondary students are likely to hold an alternative conception of the way that movement relates to force.[25, 26] It has been shown that students usually think that a continuously applied force will result in a body reaching a maximum speed, as an applied force gives an object a certain amount of 'impetus', which then somehow wears-off or dissipates. This is one of the alternative conceptions that is known to be very tenacious, and has been found to commonly recur despite instruction in Newtonian mechanics.

A common alternative conception from biology concerns the origin of the matter in plants such as trees. When people are given a piece of wood and asked how the material got into the tree they commonly reply that most of it came from the soil, although this is not the 'scientific' answer. I have seen footage of American engineering graduates and graduating science teachers in England confidently explaining that the mass of the tree came from the soil. Presumably most of these graduates would have been able to explain the basics of photosynthesis (had that been the question) and perhaps they had stored their learning about the abstract scientific process (where the carbon in the tree originates from gaseous carbon dioxide in the air) in a different compartment from their 'everyday knowledge' that plants get their nutrition from the soil.

When national UK test data was analysed by researchers for the Children's Learning in Science Project (CLiSP), it was concluded that only a third of 15 year old students used scientifically acceptable ideas about plant nutrition.[27]

Although the literature which describes alternative conceptions in science is vast, a very good (if slightly dated now) overview relating to the secondary science curriculum is available.[28]

Teachers may be able to see examples of their own students' alternative conceptions when using probes, such as **Elements, compounds and mixtures**, and others, in the companion volume to this book.

Alternative conceptions in chemistry

Chemistry is not exempt from its share of alternative conceptions. For example it has been found that it is common for post-16 students studying chemistry to think that:

- the nucleus of an atom gives rise to a certain amount of attractive force which is shared between the electrons in the atom.

This 'conservation of force' principle[29, 30] can be used (at a simple qualitative level, anyway) to 'explain' why successive ionisations require greater energy. Each time an electron is removed from an

RS•C

atom or ion the nuclear force is shared among a smaller number of electrons - so they each experience more force than before. (This is certainly an easier way of explaining the phenomena than the accepted scientific version!)

This is one example of where ideas from physics are used to explain aspects of chemistry, and it might be thought that students studying both subjects would not be likely to hold this alternative conception. However, it seems some students store their physics and chemistry learning in different memory domains, and do not easily apply their understanding about forces and electrostatics in chemistry.[31]

There are many other examples of alternative conceptions relating to chemistry topics. The Royal Society of Chemistry has commissioned a report on alternative conceptions in chemistry, which is freely available for consultation or downloading on the Internet.[32]

Consider the following examples:

■ a neutralisation process always produces a neutral product;[33]

■ in a nucleus the neutrons have the job of neutralising the charge on the protons;[34]

■ isomers are always members of the same class of compounds (*eg* both alcohols, but not an alcohol and an ether);[35]

■ a hydrogen bond is a covalent bond to hydrogen.[36]

The first example shows the importance of language in learning science. Hans-Jürgen Schmidt believes the label ('neutralisation') suggests to students that the process should give a neutral product. (The examples students meet early in their school chemistry usually do, which reinforces the idea!) A similar effect may explain why students often expect all freezing temperatures to be experienced as cold, and all melting temperatures (even for the same substances) to always be experienced as hot.[37] This effect could also be important in the second example where students are taught that a nucleus contains positive charges (which they should know will repel each other) and neutrons. The natural tendency to look for an explanation, and the suggestive label, might explain why students hold this alternative conception.

The third example is an example of learners applying the wrong level of generalisation. Students appreciate some key aspects of what isomers are, but restrict the application of the idea to within a single class of compounds. This is not a difficult conception for teachers to tackle, as long as they are able to diagnose it.

The fourth alternative conception in the list refers to an error in categorisation, with the term 'hydrogen bond' taken to mean a covalent bond to hydrogen, rather than a type of intermolecular bond. It seems this idea sometimes arises because students meet hydrogen bonds in biology in the context of nucleic acids and proteins, before they have studied this type of bonding in chemistry (see Chapters 7 and 10). If some teachers simply label a bond as a hydrogen bond without being clear what this means, it is not surprising that students' attempts to make sense of the information in terms of their existing knowledge may lead them to assuming the bond is covalent.

Students commencing post-16 science courses often only have any detailed knowledge of two types of chemical bond – ionic and covalent – so in the absence of any charges being shown, any bond drawn as a line is likely to be identified as a covalent bond.

Now this explanation may seem to suggest that some biology teachers are being careless in not making it clear that a hydrogen bond is a particular type of bond. Yet the research evidence suggests that many post-16 students have great difficulty in learning about new types of bonding. Although classing a hydrogen bond as a covalent bond can be considered as an alternative conception, it may also be part of a more complex alternative framework – the octet framework – for thinking about chemical bonding, that has been found to be quite common by the time a student leaves school.

RS•C

An example of an alternative framework

The octet framework describes the way many students make sense of school ideas about a number of aspects of chemistry (see Figure 1.1).[38, 39] For those who go on to study the subject at post-16 level, these ideas then influence how they make sense of the chemistry they are taught. Although no two students have exactly the same set of ideas, the octet framework describes aspects of student thinking that have been found to be very common. (If you are teaching post-16 chemistry you will probably find most of your students share at least some aspects of this way of thinking.)

Without describing all the evidence for this framework in detail, the following overview shows how this is not just a set of unrelated alternative conceptions. Some aspects of the framework are quite close to scientific thinking, and others may appear quite bizarre – but it is the way these ideas can be integrated into a coherent scheme that is so significant (see Chapter 10).

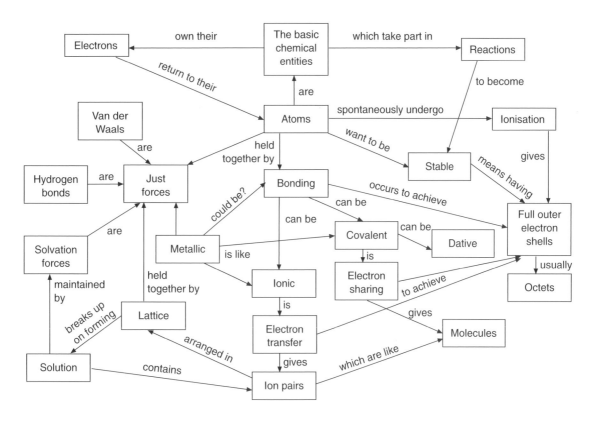

Figure 1.1 An alternative conceptual framework [40]

This framework appears to develop from learners' attempts to understand why bonds form, and why reactions occur. According to this alternative framework:

- reactions occur between atoms (not ions, molecules, lattices etc);

- the reactions occur, and bonds form, so that atoms can obtain full outer shells (or octets) of electrons; and

- there are two ways that atoms can obtain full shells: by electron transfer (ionic bond) or electron sharing (covalent bonding).

Students often take the idea that 'everything is made of atoms' too seriously (see Chapter 6). For one thing they take the meaning that everything is made from atoms to mean that the reactants in chemical reactions always start off as atoms. So they think that sodium chloride is made from atoms

RS•C

of sodium which donate electrons to atoms of chlorine - even though this is not a very likely process chemically! As the ionic bond is identified with the process of electron transfer, each ion is only considered to be fully bonded to one counter ion (see Chapter 8).

Atoms are considered to maintain their discrete identity within molecules, so that bonding electrons are still considered to belong to (and be part of) the atom from which they originated. Students therefore expect atoms to reclaim 'their own' electrons when the bond breaks (making it difficult for students to appreciate heterolytic bond fission). Some students even think that ions must be discharged by electrons returning to 'their own' atoms before new ionic compounds can be precipitated (see Chapter 9).

Because students usually fail to appreciate the physical (that is, electrical) basis of bonding they explain chemical change in anthropomorphic terms: that atoms 'want' or 'need' to have full shells (see Chapter 6). Of course there are very few chemical processes where reactants consist of atoms with partially filled electron shells - so this only makes sense because students tend to think of chemical processes as starting with atoms (see Chapter 9).

As bonding is understood in terms of the need to obtain octets or full shells, students find difficulty in making sense of bonding that can not clearly be seen in these terms. Students will have learnt two mechanisms by which they think full shells can be achieved (electron transfer and electron sharing). So metallic bonding is often seen (initially) as being like ionic and/or covalent bonds. Polar bonding is usually seen as a type of covalent bond (rather than being something intermediate between covalent and ionic). As we saw above, hydrogen bonding may be simply assumed to be a type of covalent bond. When it becomes clear to the student that this is not what is meant, it is then likely to be dismissed as 'just' a force, and not a real chemical bond (see Chapter 8).

Students will have learnt about the stability of electronic structures isoelectronic with the noble gases in their school science. Unfortunately they often generalise this idea beyond the point at which it is scientifically appropriate. Many students at this level consider an isolated sodium cation to be more stable than the isolated atom, and assume than the atom will spontaneously emit an electron, but that the positive cation could not spontaneously attract a negative electron. Some students (even after having studied patterns in successive ionisation energies) will claim that only one electron can be removed from the sodium atom - as it then has an octet.

Any reader who doubts how common these ideas are might try some of the relevant diagnostic probes included in this resource. Would you expect your students to tell you that a highly charged anion of a metallic element (Na^{7-}) is more stable than the neutral atom? The evidence from published research, and the experiences of teachers trying out these resources for the RSC, suggest that many of your students will argue that the sodium anion is more stable as it has a full [sic] outer shell of electrons! (See Chapter 6).

Hopefully, reading through this section has persuaded you that students' alternative conceptions in chemistry should be taken seriously. This publication has been designed to help you tackle this issue in the classroom. In this resource and its companion volume you will find:

■ information about some of the key alternative conceptions that have been uncovered by research;

■ copies of probes you can use to identify these ideas among your own students;

■ some specific exercises aimed at challenging some of these alternative ideas;

■ ideas about teaching approaches that may help avoid students acquiring some common alternative conceptions;

■ general ideas about helping your students develop appropriate scientific conceptions; and

■ examples of classroom activities that will help students construct the chemical concepts required in the curriculum.

Notes and references for Chapter 1

1. D. Edwards & N. Mercer, *Common Knowledge: The Development of Understanding in the Classroom*, London: Routledge, 1987.

2. K. S. Taber & M. Watts, Learners' explanations for chemical phenomena, *Chemical Education: Research and Practice in Europe*, 2000, **1** (3), 329–353, available at **http://www.uoi.gr/cerp/** or **http://www.rsc.org/Education/CERP/index.asp** (accessed September 2005).

3. J. Piaget, *The Child's Conception of The World*, St. Albans, UK: Granada, 1973.

4. Many useful references to the literature are given in the review article: K. S. Taber, Chemistry lessons for universities?: a review of constructivist ideas, *University Chemistry Education*, 2000, **4** (2), 26–35.

5. Although learners' ideas in science have attracted considerable attention, alternative conceptions have also been found in other areas of the curriculum. See D. P. Newton, *Teaching for Understanding*, London: RoutledgeFalmer, 2000.

6. R. Driver, *The Pupil as Scientist?* Milton Keynes: Open University Press, 1983.

7. For example: R. Driver, J. Leach, R. Millar & P. Scott, *Young People's Images of Science*, Buckingham: Open University Press, 1996.

8. H. Dickson, *Fantastic Optical Illusions and Puzzles*, London: Lagoon Books, 1996.

9. One of the teachers who piloted materials for this project reported that his group of 13–14 year olds 'were reluctant to relinquish their own ideas' when presented with the materials on Elements, compounds and mixtures.

10. G. Claxton, Minitheories: a preliminary model for learning science, in P. J. Black & A. M. Lucas, *Children's Informal Ideas in Science*, London: Routledge, 1993, 45–61

11. J. Kuiper, Student ideas of science concepts: alternative frameworks?, *International Journal of Science Education*, 1994, **16** (3), 279–292.

12. J. K. Gilbert, R. J. Osborne & P. J. Fensham, *Children's Science and its Consequences for Teaching, Science Education*, 1982, **66** (4), 623–633.

13. J. Bliss, I. Morrison & J. Ogborn, A longitudinal study of dynamics concepts, *International Journal of Science Education*, 1988, **10** (1), 99–110

14. R. Driver, *The Pupil as Scientist?* Milton Keynes: Open University Press, 1983

15. D. Dumbrill & G. Birley, Secondary school pupils' understandings of some key physics concepts, *Research in Education*, 1987, **37**, 47–59.

16. L. Viennot, Spontaneous reasoning in elementary dynamics, *European Journal of Science Education*, 1979, **1** (2), 205–222

17. L. Viennot, Analyzing students' reasoning: tendencies in interpretation, *American Journal of Physics*, 1985, **53** (5), 432–436.

18. Qualifications and Curriculum Authority (QCA), *Standards in Key Stage 3 Science (2000)*, London: QCA, 2001, 12.

19. J. Solomon, *Getting to Know about Energy - in School and Society*, London: Falmer Press, 1992.

20. G. Aikenhead, Renegotiating the culture of school science, in R. Millar, J. Leach & J. Osborne, *Improving Science Education: the Contribution of Research*, Buckingham: Open University Press, 2000.

21. P. Georghiades, *Conceptual change learning in primary science: a step forward?*, 1999, available via Education-line, at **http://www.leeds.ac.uk/educol/** (accessed September 2005).

RS•C

22. J. R. Anderson, *Learning and Memory: An Integrated Approach (2nd Edition)*, New York: John Wiley & Sons, 2000.

23. J. R. Anderson, *Cognitive Psychology and its Implications (4th Edition)*, New York: W. H. Freeman & Company, 1995.

24. I. O. Abimbola, The problem of terminology in the study of student conceptions in science, *Science Education*, 1988, **72** (2), 175–184.

25. D. M. Watts & A. Zylbersztajn, A survey of some children's ideas about force, *Physics Education*, 1981, **16** (6), 360–365.

26. J. K. Gilbert & A. Zylbersztajn, A conceptual framework for science education: The case study of force and movement, E*uropean Journal of Science Education*, 1985, **7** (2), 107–120.

27. B. Bell & A. Brook, *Aspects of Secondary Students' Understanding of Plant Nutrition: Full Report*, Leeds: Children's Learning in Science Project, Centre for Studies in Science & Mathematics Education, University of Leeds, 1984.

28. R. Driver, A. Squires, P. Rushworth & V. Wood-Robinson, *Making Sense of Secondary Science: Research into Children's Ideas*, London: Routledge, 1994.

29. K. S. Taber, The sharing-out of nuclear attraction: or I can't think about Physics in Chemistry, *International Journal of Science Education*, 1998, **20** (8), 1001–1014.

30. K. S. Taber, Ideas about ionisation energy: an instrument to diagnose common alternative conceptions, *School Science Review*, 1999, **81** (295), 97–104.

31. K. S. Taber, The sharing-out of nuclear attraction: or I can't think about Physics in Chemistry, *International Journal of Science Education*, 1998, **20** (8), 1001–1014.

32. V. Barker, *Beyond appearances: Students' Misconceptions about Basic Chemical Ideas:* London: Royal Society of Chemistry, 2000, available on LearnNet at **www.chemsoc.org/LearnNet/miscon.htm** (accessed September 2005).

33. H-J. Schmidt, A label as a hidden persuader: chemists' neutralization concept, *International Journal of Science Education*, 1991, **13** (4), 459–471.

34. H-J. Schmidt and T. Baumgärtner, *Senior high school students' concepts of isotopes – a triangulation study.* In press.

35. H-J. Schmidt, Conceptual difficulties with isomerism, *Journal of Research in Science Teaching*, 1992, **29** (9), 995–1003.

36. K. S. Taber, Building the structural concepts of chemistry: some considerations from educational research, *Chemical Education: Research and Practice in Europe*, 2001, **2** (2), 123–158, available at **http://www.uoi.gr/cerp/** or **http://www.rsc.org/Education/CERP/index.asp** (accessed September 2005).

37. A. Quinn, Conflicts in perception, in J. Sanger, *The Teaching, Handling Information and Learning Project,* Research Report 67, London: The British Library, 1989.

38. K. S. Taber, An alternative conceptual framework from chemistry education, *International Journal of Science Education*, 1998, **20** (5), 597–608.

39. K. S. Taber, Alternative conceptual frameworks in chemistry, *Education in Chemistry*, 1999, **36** (5), 135–137.

40. Diagram taken from: K. S. Taber, Molar and molecular conceptions of research into learning chemistry: towards a synthesis, 2000, available via Education-line, at **http://www.leeds.ac.uk/educol/** (accessed September 2005).

RS•C

2. Concepts in chemistry

This chapter discusses the nature of chemical concepts and how such concepts are learnt. The problem of clearly defining key chemical concepts such as 'element' and 'molecule' is explored, and the implications for teaching are considered.

What are chemical concepts – and why are they hard to learn?

The term 'concept' is used a lot when talking about learning, but it is one of those words – perhaps like 'molecule' (see below) – which although we seem to know what we mean by it, is not so easy to define precisely.

Psychologists refer to the ways in which we represent our knowledge as 'schema', and concepts (or categories) are important types of schema for making sense of our world.[1] A concept is just a way of breaking up the world into bits that we can recognise, and think about. Some of our concepts – so called identity concepts – refer to specifics such as your particular school or college, or particular people. I am writing this at a place I recognise as the University of London Institute of Education. I have a mental representation – a concept – of that institution. Other types of concept – so called equivalence concepts – do not refer to specific items, but to categories that may have several, or even many, members. It is these types of concepts that are important in learning science.

Even among these categories there are important distinctions. Some concepts are ad hoc – *ie* they are made up as we go along for a particular purpose (such as the group of students who have failed to bring the homework and who have been asked to remain at the end of today's class). More significant for our present purposes are natural concepts and rule-governed concepts.

People the world over tend to categorise certain things in the same ways. For example we all use categories such as 'tree' without having to think about how such terms are defined – we think we know what a tree is when we come across one! These concepts seem to be largely intuitive – we do not seem to need to be taught the concept. We may teach a young child the word 'tree' but somehow she will learn what is, and what is not, considered a tree without either being given a detailed definition, or having to be taken through large numbers of examples of trees and non-trees. We say that the child recognises a 'family resemblance' between objects considered as trees.[2] This is an important point that is not always recognised by teachers – although students often seem to have great difficulty in learning aspects of science regardless of well structured teaching, those same students somehow learn to use natural categories with only a minimal and often unplanned opportunistic 'teaching' input. This suggests that the brain is 'programmed' to be 'ready' for such instruction.[3]

Now this is obviously not supernatural: rather, through millennia human brains have evolved to be predisposed to recognise the types of categories which are useful to survival. Although this makes some types of learning much easier, it can also be a nuisance in formal education.

This is because, in science, we are largely dealing with rule-governed rather than natural concepts. In other words we often define new concepts.[4] Unfortunately some scientific definitions may seem to be inconsistent with natural ones.[5] For example, in science we have a concept of 'animal' which includes a wide range of living things such as various types of worms, fish, birds, insects etc. Yet there is also a natural category of living things limited to large mammals such as horses, cows, pigs, deer etc (and not including humans) which are commonly described as 'animals' in everyday life. Insects are not considered animals in this sense. Spiders – which according to science are animals and are not insects – are usually considered as insects, but not animals.

It is difficult to overcome this problem for a number of reasons. The natural category meanings are usually learnt before the scientific ones, and are very commonly used in everyday life (and so get reinforced regularly). Their 'lifeworld' (everyday) meanings for words such as 'animal' are useful in

RS•C

many contexts, therefore it is difficult to expect students to give them up. Also they are different types of concepts – and so they are difficult to fully reconcile. Natural categories are not defined by any 'hard and fast' rules, and may have very fuzzy boundaries (so today my hamster will be an animal as we are talking about animals as pets, but tomorrow when we talk about animals as animals in the wild...), whereas in science we (usually) use rules to define our concepts.

In chemistry these problems may seen less extreme than in biology, or physics (where most of the key concepts are blighted by having labels which are also used in more vague everyday senses: 'energy', 'momentum', 'force', 'impetus', 'work'...[6]), but we may still find students using words like 'natural', 'synthetic', 'metal' 'pure' and 'substance' in ways which do not match the technical meanings. [7]

We might expect that learning about chemical concepts which do not have everyday counterparts – such as electron, reducing agent and hybridised atomic orbital – will not trigger the same problems. Yet we all know that students do have difficulties acquiring such concepts. Perhaps the problem is that the structure of our brains has evolved to be inherently good at learning natural concepts (by a kind of cultural osmosis with feedback), but are less well suited (adapted) to learning the type of rule-based concepts so important in science.

Forming concepts – noticing similarities and differences

In one sense a person can be said to have acquired a concept once they are able to identify examples, and distinguish them from non-examples. This may be the case whether the concept concerned is that of 'cat' or 'cation'. Acquiring and using concepts therefore requires the learner to recognise similarities and differences.

For example, consider the two figures below (Figure 2.1 and Figure 2.2).

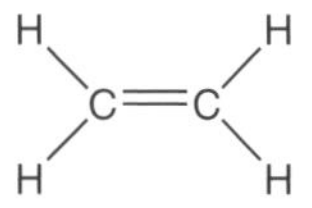

Figure 2.1 A molecule

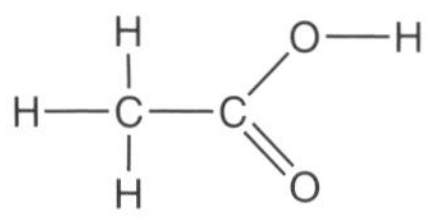

Figure 2.2 A (different) molecule

A student may recognise that they are both examples of molecules (if they have a concept of molecule), and perhaps that they are both molecules of substances considered organic (if...). The student might identify that both compounds include a double bond (if...). Perhaps the student could identify the first as an example of a hydrocarbon (if...), and further as an alkene (if...) and even as ethene (if...); and the second as a carboxylic acid (if...), and perhaps as ethanoic acid (if...). Making these classifications requires the students to recognise certain attributes of the species represented as pertinent to these various concepts.

One way to focus students' thinking on the relevant attributes is to ask them to spot similarities and differences (*eg* between diagrams such as those above). An approach that has been used in educational research is to present learners with three diagrams, and ask them to (a) suggest which is the odd-one-out and (b) explain why. This method requires the student to make discriminations between the diagrams, and so elicits the features (the 'constructs') they bring to mind when thinking

RS•C

about the diagrams. This approach, known as 'Kelly's triads' can produce some interesting suggestions, and may reveal both areas of ignorance (when students fail to spot 'obvious' chemical similarities and differences) and alternative conceptions.

Such an approach can allow the teacher to explore the student's inventory of chemical constructs (by working through a series of comparisons with different 'triads' of diagrams). When this techniques was used with some post-16 students who were presented with triads showing various combinations of molecules, atoms, ions etc, it was possible to develop a simple framework for the types of comparisons these students made.[8]

The four main classes of suggestions from students concerned aspects of the structure of the chemical species; comments about the properties of the species or the substance it was a component part of; classification of the species (or the substance it was a component part of) in various chemical categories; and aspects of the way the species was represented in the particular diagram used.

By matching the constructs used by individual students to this framework it is possible to get an impression of the types of features that a particular student focuses on.

Consider the set of constructs elicited from two classmates, Kabul and Rhea, when they undertook the exercise during their first term of post-16 chemistry. (In these schemes the normal text represents the framework developed from all the responses, and the bold print shows the constructs used by the individual students). Kabul, who went on to be very successful in the course, uses a range of ways of discriminating chemical species from the four main categories.

RS•C

Kabul's constructs:

structural:
  molecular:
    shape: **tetrahedral arrangement**
    others:
  sub-atomic:
    nuclear:
    electronic:
      c.f. noble gas electronic configuration:
        possess octet state
      others:
  crystal: **lattice arrangement**
  bond type: **covalent bonding; bond between different elements; ionic compound; bond between non-metals; polar covalent bond**
  includes:
properties:
  chemical:
    reactivity: **can undergo reaction; can undergo reaction to form ionic bonds; cannot exist on its own; high reactivity; stable**
    specific: **forms diatoms; displacement of hydrogen by reactive metals; can undergo combustion**
    valency: **electrovalency of -2; covalency of 4; electrovalency of 1**
  physical:
    macroscopic: **low melting point; soluble in organic solvents; conduction of electricity; soluble in water**
    molecular: **high energy required to break bonds**
      charge: **charged particle; a gain of electrons; ionising slowly**
      others:
  environmental:
classification:
  periodic table:
    electronegativity: **metal**
    block:
    period:
    group: **found in group 7; found in group 1; found in group 8**
  state: **state of existence is solid**
  reagent type:
  microscopic species: **represents an ion**
  type of substance: **only one element; organic substance; compound**
  specific substance:
  occurrence:
diagrammatic features: **we can know the period; represents a type of bond**
ambiguous/miscellaneous: **can be present in a noble gas; ionisation**

However, classmate Rhea made many fewer chemically significant discriminations, and indeed often seemed to largely focus on such graphical conventions as whether electrons were represented as 'e', 'e⁻' or '•'. Rhea later decided not to continue with her study of chemistry, but to concentrate on her other academic subjects.

Rhea's constructs:

    structural:
        molecular:
            shape: **symmetrical-ish; circular**
            others: **two joined together; all clumped together**
        sub-atomic:
            nuclear: **got a 17+ charge in the middle**
            electronic:
                c.f. noble gas electronic configuration:
                one electron short of a full outer
                shell; full outer shell
                others: **three shells**
        crystal:
        bond type: **double bonds drawn in**
        includes: **got orbitals; got 'H's; two different elements in them**
    properties:
        chemical:
            reactivity:
            specific:
            valency:
        physical:
            macroscopic:
            molecular:
                charge:
                others:
            environmental:
    classification:
        periodic table:
            electronegativity:
            block:
            period:
            group:
        state:
        reagent type:
        microscopic species:
        type of substance:
        specific substance:
        occurrence:
    diagrammatic features: **other shells drawn in ; electrons as circles; electrons as**
        'e's; say what they are; minus signs on some of the 'e's; got shading;
        got brackets; written; got plus signs; say how many electrons are
        shared; got plus signs in the middle; got charges drawn in; 3-D
        drawing; simple sketch drawing; got a key
    ambiguous/miscellaneous: **got structure(s)**

The Kelly's triads technique is usually used on a one-to-one basis, but can be adapted for class use. An alternative approach for the classroom, however, is to ask students to suggest similarities and differences between two diagrams (or other suitable stimuli).[9]

One of the resources included in this publication is a set of probes using such dyads (pairs) of pictures, **Chemical comparisons**. There is an almost unlimited possibility for developing questions with various dyads, and so the specific probes included should be seen as exemplars.

Some of these examples are suitable for use with 11–14 year olds. For example Figures 2.3 and 2.4 show cylinders labelled as iron and sulfur.

RS•C

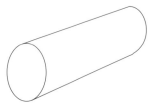

Iron

Figure 2.3 A picture of iron presented to students

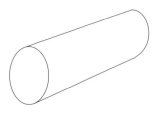

Sulfur

Figure 2.4 A picture of sulfur presented to students

Responses to such dyads are able to demonstrate significant differences between the repertoires of ideas that students call upon in making comparisons. One student in a group of 13–14 year olds suggested four similarities in this example: 'they are both solids', 'they are both elements', 'they are both in rods' and 'they are the same size'. This student indicated that the former two suggestions, but not the latter two were important to chemists. The same student suggested five differences: iron being a metal, colour, iron being malleable, poor versus good conductor, iron being ductile (and she thought that all but colour were significant differences). On this same question another student in the group was only able to suggest that 'they are both the same shape & size'.

Within the group of students there were a wide range of suggestions, many of which were valid (see Table 2.1). Some of the suggestions students make can reveal alternative conceptions (such as believing iron is found native), and the sophistication of answers can vary. However, this latter point should not be considered as a draw-back. As an open-ended activity this technique allows 'differentiation by outcome'. In other words, lower attainers should be able to contribute ideas, whilst there is scope for the most able to think up suggestions related to a wide range of relevant chemical themes.

RS•C

	Similarities	Differences
Acceptable	Same shape/cylindrical/rods Same size Elements Both appear on the Periodic Table Both combine with other things Both react with oxygen Solid at room temperature	Conductivity (electrical) Conductivity (thermal) Density Weight Boiling temperature Melting temperature Metal/non-metal Magnetic Silver-grey/ yellow Hard/powdery Malleability Ductility Ease of cutting up Iron rusts Feel different 'Shineness'
Dubious	Colour Both natural solids (minerals) Don't look malleable Have the same molecules set-up Appear to have the same mass Exactly same smoothness	Sulfur is more reactive

Table 2.1 Comparisons between iron-sulfur dyad by students in one class

Many of the probes included in the resource use diagrams of molecular level systems, rather than diagrams of macroscopic samples. For example, Figures 2.5 and 2.6 show a dyad representing solid and aqueous sodium chloride.

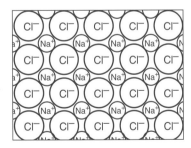

Figure 2.5 A representation of solid sodium chloride NaCl presented to students

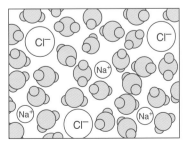

Figure 2.6 A representation of sodium chloride solution presented to students

RS•C

This dyad was intended primarily for students in the 14–16 age range, but was attempted by some students in the group of 13–14 year olds. Their suggestions are given in Table 2.2. Although they made some valid observations (about the balance between positive and negative charge, and the difference in the order of the arrangements), there was also evidence of some confusion over basic ideas. For example, it was suggested that the ionic solid was a mixture, and that the solution was a gas (perhaps because the ions themselves appeared quite spread out). Clearly this exercise can lead to useful discussion points.

	Similarities	Differences
Acceptable	Both contain Na^+ and Cl^- Same amount of Na^+ and Cl^- Same amount of positive and negative ions	Not joined similarly Evenly spread – randomly spread Compound – mixture
Dubious	Both represent mixtures They're both compounds Same substance in different states	One is a solid and one is a gas

Table 2.2 Comparisons between solid and aqueous sodium chloride made by students in one class

As students learn more chemistry it is possible to ask them to make more complex or subtle comparisons. So many of the probes included in this resource are intended primarily for students on post-16 courses. The exercise was carried out with a group of 17–18 year olds nearing the end of their chemistry course. These students were able to suggest many comparisons that would not have been possible for the younger students. For example, one student compared a dyad representing NH_3 and BCl_3 (Figures 2.7 and 2.8).

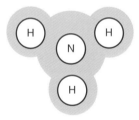

Figure 2.7 A representation of an ammonia molecule

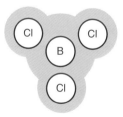

Figure 2.8 A representation of a boron trichloride molecule

The student identified a number of valid differences between these superficially similar species: tendency to dimerise (BCl_3), state (gas – solid), electron deficient nature (BCl_3) and bonding character. However, in the last case the bonds were described as covalent (NH_3) and ionic (BCl_3), suggesting

RS•C

that the student should be encouraged to think in terms of the extent of bond polarity rather than in absolute terms (see Chapter 8).

The examples discussed here have shown how the simple technique of asking students to spot similarities and differences can be useful for auditing knowledge and spotting alternative conceptions, as well as providing some variety through an open-ended activity. A similar approach may be used to explore analogies used in teaching. When introducing difficult ideas, which do not have any obvious 'anchors' into students' existing experiences, it can be useful to use analogies to help students make sense of the unfamiliar material (see Chapter 4). However, analogies can also lead to inappropriate and wrong learning unless students are very clear about which features of the analogue they are meant to adopt (see Chapters 7 and 10).

One example of a teaching analogy that is sometimes used is that of the atom being like a tiny solar system. Although this can be useful, it can also lead to students making incorrect assumptions about atoms (either due to a poor understanding of solar systems, or due to an over-zealous adoption of the comparison). The approach used in the **Chemical comparisons** resource has also been adopted in another probe to explore whether students appreciate the atom-solar system analogy. The details of that probe, and how students respond to it, are given in Chapter 7, but it is worth noting here that teachers could adopt the approach used here when using their own teaching analogies or if concerned about the analogies that students may bring to class. The **Chemical comparisons** probes may be readily adapted by replacing the diagrams with any others the teacher may wish to use – for example, pictures of the π-cloud in benzene and a doughnut – if this is an analogy used with students.

The technique of focusing on similarities and differences can also be useful in responding to specific problems students may have in making scientific discriminations. One learning difficulty that has been reported in chemistry is that of not distinguishing between strong acids and concentrated acids. It is understandable that students might see these terms as synonymous, as in everyday life a concentrated solution is often described as a 'strong' solution, whereas in chemistry acidic solutions may be weak and concentrated or strong and dilute (as well as strong and concentrated and weak and dilute). Included in the companion volume is a classroom probe, **Explaining acid strength**, which allows teachers to diagnose whether their students are clear about the distinction between acid strength and concentration. This is accompanied by a classroom exercise, **Classifying acid solutions**, which presents a set of diagrams to help support students making the discrimination. The exercise presents diagrams of acid solutions which vary along the two dimensions (strong-weak and dilute-concentrated), and students are encouraged to focus on the pertinent attributes when comparing the figures (ie the amount of solute present in the solution, and the presence or absence of associated solute molecules as well as ions). In this way they are led to classify each diagram as strong/weak and dilute/concentrated. When these materials were piloted for the project it was found that students were generally comfortable in applying the ideas, and that those who had been unclear found the exercises helpful.

Learning rule-based concepts

In principle it might sound easy to teach rule-based concepts. A sensible approach would be:

- Provide the learner with the rules for, or the definition of, the concept;

- Give a few salient examples, and non-examples, and make sure the learner understands why they are, or are not, examples of the concept;

- Provide practice exercises with plenty of examples and non-examples; and

- Require the learner to be explicit about their reasoning in working through the exercises.

Although such an approach would seem a very sensible way of making sure that students have grasped important new concepts you introduce, it is not foolproof.

RS•C

For one thing, many concepts are more complicated than they may at first appear. This means that making the 'rules' explicit can get tricky. It may be difficult to provide definitions that are both comprehensive and accurate. Definitions that meet these requirements, and are also accessible to learners may be very hard to find. (Some examples of definitions of basic chemical terms are discussed below.)

A second complication is that our definitions use language, and this introduces two more problems. Some students are not easily able to understand complicated sentences.[10] Even when this is not an issue, many of the words used in a definition are themselves labels for other concepts that also need to be understood. If we were to define an alkene as a type of hydrocarbon with a double bond we are assuming that the learner already has an acceptable understanding of both the concepts of 'hydrocarbon' and 'double bond'.

Providing clear definitions

Unfortunately, many chemical concepts are only 'simple' once they are understood as part of a much wider network of ideas. (The importance of appreciating the way concepts are related is discussed in Chapter 3.) So consider the following definition:

An alcohol is a compound with an –OH group.

This definition would include ethanol etc, but could also apply to ethanoic acid, and even sulfuric acid – which are not considered alcohols. So we would need to add something about the alcohol being an *organic* compound and the –OH group not being part of a larger –COOH group (or an –OOH group for that matter).

If we defined an alkene as having an empirical formula of C_nH_{2n}, then we would need to add the proviso that it was not a cyclic molecule (in which case C_nH_{2n} would be an alkane, and the alkene would be C_nH_{2n-2}).

Sometimes we can avoid such complications by only providing a limited definition initially, because we only wish to use the definition (and therefore the concept) in a limited range of cases. If learners do not yet know about the existence of cyclic compounds, then they should not misapply the C_nH_{2n} = alkene rule to cyclic alkanes. However, even here we may be storing up problems for the future if we are not careful. For example, if we were to define oxidation as 'the addition of oxygen', we may later wish to add a new 'or the removal of hydrogen', and then later add 'or the loss of an electron', and then even later add 'or an increase in oxidation state'.

There is a real issue here of finding the optimum level of simplification, of balancing the need to keep things simple now, and providing learners with ideas which we are not going to expect them to alter later.[11] Learners do not always find it easy to adjust meanings for technical words once they are acquired. It could be argued that science is largely about culturally agreed models that scientists use to make sense of the world.[12] Moreover, scientists see models as tools: they are usually limited (which means we have to sometimes switch to a different tool), and they are for our benefit (which means we are able to play with them, and even change or discard them). Yet students tends to see our scientific models as simple descriptions of the way nature is. Even when scientific ideas are seen as models, we must remember that – for most students – models are expected to be incomplete copies of nature – rather than convenient abstractions (see Chapter 6).

As teachers we should remember that although there are things in the world which are (by our definitions) oxidising agents or alkalis – the categories themselves are artificial and just for our convenience. For example, acids occur in nature – the idea of 'acid' only exists in our minds. We should also remember that most of these concepts are recent human inventions, and their meanings have changed during the development of science. If we can encourage our students to understand this, then they may find it less stressful when we suddenly expect them to accept bonds that are 'in-between' covalent and ionic (see Chapter 8), or an acid that does not contain hydrogen (eg SbF_5).

RS•C

Defining key concepts in chemistry

It is useful to illustrate the problems associated with defining chemical concepts, and so we will consider a small number of the key concepts in chemistry as examples.

Elements and compounds: two of the central concepts in chemistry (see Chapter 6) are 'element' and 'compound'.

Atoms and molecules: the importance of the particle model to making sense of chemistry is well recognised (see Chapter 6), and the concepts of 'atom' and 'molecule' are widely used in the subject at all levels.

Chemical and physical changes: chemistry is largely concerned with the properties and reactions of substances. Properties are often classed as physical (colour, melting temperature...) and chemical (which substances are reacted with, under what conditions, to give which products). Reactions, chemical changes, are a central part of the subject (see Chapter 9). An understanding and application of the distinction between physical and chemical changes is often expected of students at lower secondary level. [13]

It is reasonable to suggest that;

(a) it is difficult to form a detailed appreciation of chemistry without a fair understanding of what is meant by these terms; and

(b) practising chemists, and chemistry teachers would have a clear understanding of what these ideas mean.

It might be expected, therefore, that school textbooks, and other reference books, would provide consistent, unproblematic, definitions for such key terms (element, compound, atom, molecule, physical and chemical change) and that chemistry teachers would tend to agree on these definitions. However, in practice, this may not be so.

Elements and compounds

Consider two definitions of the term 'element' from chemistry textbooks aimed at the secondary level:

'An element is a substance which cannot be split up into simpler substances.'[14]

'A substance that is made of only one kind of atom.' [15]

These are very different definitions, one working at the macroscopic level and one at the molecular level (see Chapter 6), but both are common ways of defining the element concept. To the chemist it is clear how these definitions relate to each other, but to a novice they may seem to have little in common. Similarly, the following definitions of 'compound' also seem to be quite distinct:

'A substance consisting of atoms of different elements joined together.'[16]

'A chemical substance made up of two or more elements bonded together, so that they cannot be separated by physical means.' [17]

'A product which has properties different from those of either of the component substances and which is formed with an accompanying energy change is called a compound.' [18]

It is valuable to consider the usefulness of these concepts, to see what lessons can be drawn for teaching using definitions of such central ideas. A classroom exercise, **Definitions**, to explore student reactions to definitions of basic terms is included in the companion volume. Students are asked about the accuracy and usefulness of a set of definitions of key terms (see Chapter 6). Some practising science teachers were also asked about these definitions, and some of their comments are quite illuminating.[19]

RS•C

An element is a substance which cannot be split up into simpler substances.

This definition requires the learner to appreciate the chemist's meaning of substance, and also to share what is meant by 'simpler' in this context. One could object that 'simple' here means an element compared with the more complex 'compounds' (and so the definition is circular).

One teacher considered this definition was 'wrong', but still 'helpful' for 13 year olds 'when they have not been taught' about protons, neutrons and electrons. Another teacher judged this definition as 'correct' but 'not helpful' as it was 'too vague'.

A teacher who thought the definition was correct explained that 'anything simpler would not be a substance'.

Another teacher thought this definition was 'wrong' and 'not helpful' as it 'needs to have 'by chemical means' added [as] CERN etc split atoms into more fundamental particles'. Yet this seems to confuse the macroscopic and molecular levels of analysis (see Chapter 6) as atoms do not directly relate to substances.

[An element is] a substance that is made of only one kind of atom.

This sounds quite straightforward, but even if students understand about atoms, they also need to appreciate the specific way in which 'one kind' is used – so that atoms with the same mass number but different atomic number of are different kinds, but atoms with the same atomic number and different mass number count as being of the same kind in this context. One teacher judged this definition as both 'correct' and 'wrong': considering it correct until 'you introduce isotopes', but thought this was a helpful definition 'at Year 10 level' [*ie* about 14 years of age]. Another teacher judged this definition as 'correct' and 'helpful' as 'it's simple and straightforward – you can also draw it & represent it visually'. Another thought 'the definition is only really useful when accompanied with plenty of examples'.

[A compound is] a substance consisting of atoms of different elements joined together.

One objection to this definition is that it would not include ionic materials, which do not consist of atoms (see Chapter 7). As one teacher commented, this is a 'common definition but causes problems in ionic bonding'. Yet some teachers felt this was an accurate definition. One teacher, who thought this was both 'correct' and 'helpful' thought it was 'easy to represent diagramatically'. Another teacher suggested that this was correct, although not necessarily helpful as a 'knowledge of 'atom' and 'element' [was] required' before it could be applied.

One teacher suggested that the accuracy of the definition depended upon how 'join' was defined, and another who also felt this was 'too vague', and should be 'chemically combined' or 'chemically bonded'.

[A compound is] a chemical substance made up of two or more elements bonded together, so that they cannot be separated by physical means.

This definition clearly requires students to already understand the idea of element, and to appreciate what is meant by 'physical means' – it 'depends upon pupils' understanding of physical & chemical change' – (see below) and 'bonded'.

One teacher thought this was a 'correct' definition, but 'not helpful' as it had 'too much detail'. Another teacher thought this was a 'helpful' definition, but 'only once [the] concept [was] learnt' as there were 'too many ideas to combine otherwise'.

[A compound is] a product which has properties different from those of either of the component substances and which is formed with an accompanying energy change is called a compound.

This rather complex definition is difficult to apply, as was pointed out by one teacher who judged it as both 'wrong' and 'not helpful', and gave the example of the reaction between copper oxide and hydrogen: copper is a product with different properties, but is 'not a compound'. Of course, it is open

RS•C

to interpretation (*ie* definition) whether hydrogen and copper oxide are 'component substances' of copper.

In any case the student would already need to understand the ideas before the definition could be useful (it was 'good for redefining knowledge when pupils have fairly secure ideas' according to one teacher).

Another teacher thought that this definition was 'wrong' and 'not helpful',

'not sure but feel it would apply to alloys and to products where components are not elements'

Another teacher suggested that this definition was wrong as 'compounds are not necessarily different from the component substances', but unfortunately did not provide any examples to support this view.

Atoms and molecules

In a similar way to our consideration of definitions of elements and compounds, we may consider how the concepts 'atom' and 'molecule' are defined in texts. Some examples of definitions of 'atom' are:

'The smallest part of an element which can exist as a stable entity.' [20]

'[Even though an atom is made up of smaller particles, as we will see shortly, it is still regarded as] 'the smallest particle of an element that still shows the chemical properties of the element.'...'[21]

'The smallest portion of an element that can take part in a chemical reaction.'[22]

'Atoms are the smallest particles that can be obtained by chemical means.'[23]

The following definitions of 'molecule' were found in common reference books:

'The smallest particle of matter which can exist in a free state' [24, 25]

'The smallest portion of a substance capable of existing independently and retaining the properties of the original substance.' [26]

'group of two or more atoms bonded together. A molecule of an element consists of one or more like atoms; a molecule of a compound consists of two or more different atoms bonded together.' [27]

[monatomic molecule] 'A molecule of an element, consisting of a single atom of an element. *eg* the molecules of the inert gases.' [28]

Again the comments of colleagues are interesting:

[An atom is] the smallest part of an element which can exist as a stable entity.

This definition is rather dubious, as it is not clear what is meant by 'stable'. One teacher pointed out that 'radioactive isotopes are not stable'. Most atoms certainly can be stable, under certain conditions (see Chapter 6). Under normal conditions (such as in a school laboratory) most discrete atoms would not be stable, and would readily interact to form molecules, ions etc. This view was not always supported by teachers considering the definitions.

One teacher judged this a 'correct' and 'helpful' definition, adding the comment: 'no problem'. Another teacher was 'not sure' if the definition was correct, asking 'how stable are protons and neutrons'. This is an interesting point, as a proton is stable in the sense of not undergoing radioactive decay: but, like most atoms, would not remain 'free' for long in most chemical environments[29]:

The word 'entity' in this definition was considered too difficult by one teacher.

[An atom is] the smallest particle of an element that still shows the chemical properties of the element.

RS•C

This is a very dubious definition. Firstly, it is difficult to know how chemical properties are defined for individual atoms. Under usual laboratory conditions the substance sodium reacts with water, the substance sulfur does not: it is difficult to know how one would judge the 'reactivity' with water of an atom of either element. (The term 'react' is not really applicable in this context – see Chapter 6.) Secondly, if one did try to evaluate the 'reactions' of the individual atoms, then atoms of substances would 'react' in conditions where the substance itself would not. An atom of carbon is going to be a 'reactive' species in chemical environments where bulk carbon is not. Again, teachers do not always take this viewpoint.

One teacher who thought that this definition was both 'correct' and 'helpful' described it as 'unproblematic'; and another thought it 'helps define 'stable entity' better to a younger pupil' and was 'OK for A level [ie post-16 level]'.

One teacher thought this definition was 'correct' but 'not helpful' as,

'This is a very theoretical type of definition. Testing it out would be impossible.'

A teacher who was unsure of whether the definition was correct posed the question 'how can an atom show properties of [the] element[?]', commenting that 'things like density etc are 'bulk' properties'. One teacher who judged this definition as 'wrong' because a 'Cl atom does not have [the] same properties as Cl_2', nevertheless thought it was 'helpful' to learners.

[Atoms are] the smallest portion of an element that can take part in a chemical reaction.

The biggest problem with this definition is that chemical reactions are macroscopic phenomena, and one should not confuse macroscopic and microscopic parts of explanations (see Chapter 6). In this case we find that such a confusion is very unhelpful. For example, if oxygen reacts, then it is certainly reasonable to say that the molecules of oxygen take part in the process. However, we could say that the atoms of oxygen take part in the reaction – and this could seem just as true. But it could also be said that the outer shell (valence) electrons take part in the reaction (which means they could be seen as atoms by this definition). This point was recognised by one teacher who argued that 'electrons take part in reactions', but other teachers found this definition 'correct' and 'helpful'.

Atoms are the smallest particles that can be obtained by chemical means.

This definition requires the learner to understand what is meant by the term 'chemical means' (see below) – as one teacher asked: 'is radioactivity chemical means?' – but it also seems dubious as most chemical process give a product which is molecular, or ionic, etc, and not a product which is atomic. Teachers asked about this definition disagreed about whether it was correct or not, and a number reported that they were unsure.

[A molecule is] the smallest particle of matter which can exist in a free state

For this to be useful the learner would need to understand what is meant by a free state. Even given this, some teachers would object to the definition. One teacher judged this as both 'wrong' and 'not helpful' because 'noble gases, metal vapour etc are monatomic'; another thought that the definition was correct, 'but causes problems for students with monatomic molecules'. As is discussed below, many teachers class monatomic molecules as atoms, but not as molecules.

The reference to 'the smallest particle of matter...' also causes problems, as one teacher suggested 'substance' might have been a more appropriate term here. Another teacher felt this was 'wrong' as it 'would confuse atoms with molecules'. A teacher who was unsure if the definition was correct suggested that 'if it's true it could be useful'!

[A molecule is] the smallest portion of a substance capable of existing independently and retaining the properties of the original substance.

This is very similar to some of the definitions of atom given above. The first section of this statement seems less dubious in this context as molecules do commonly exist independently (where atoms do not).[30] The second part of the definition is more problematic. Molecules of a substance can be

RS•C

considered to show some of the properties of a substance (sometimes colour, smell), but, again, can not really be considered to have bulk properties such as hardness, density or conductivity. As one teacher (who judged the definition to be 'wrong') explained;

'a single molecule can't be solid, liquid or gas as it is the interaction with others that give the state of the matter'.

One teacher judged this definition as 'correct' and 'helpful' and 'unproblematic'. However, other teachers who also considered this definition as 'correct' judged it as 'not helpful' as it was 'too complicated & subtle' and the language was 'too difficult' for learners.

[A molecule is a] group of two or more atoms bonded together. A molecule of an element consists of one or more like atoms; a molecule of a compound consists of two or more different atoms bonded together.

This definition is not self consistent, as the second sentence, allowing the possibility of monatomic molecules, contradicts the first. Yet some teachers considered it both 'correct' and 'helpful': describing it as a 'good definition at KS4 [*ie* for 14–16 year old students]'; 'useful at end [of teaching the topic]', and noting that it 'distinguishes between elements and compounds'. One teacher who thought it was 'wrong', nevertheless thought it was 'helpful' and the 'definition most people work with'.

[A monatomic molecule is] a molecule of an element, consisting of a single atom of an element. Eg the molecules of the inert gases.

In my own undergraduate study I was quite familiar with the notion of monatomic molecules, as opposed to diatomic or polyatomic molecules. One teacher who thought that this definition was 'correct' thought that it was 'helpful' as 'the definition gives an example'. Yet, I am aware that many chemistry teachers do not feel that a single atom should be described as a molecule, even when the atom is stable, and is the smallest part of the element which may be considered to share some of its properties (*ie* the noble gases).[31] One teacher, classing this definition as both 'wrong' and 'not helpful' explained that 'I don't think Ar is a molecule!' Another was more insistent that a 'single atom is not [a] molecule'. One teacher who was unsure about the accuracy of the definition explained,

'I think 'monatomic molecule' is a contradiction in terms.'

Clearly if teachers themselves consider that a molecule is always a group of atoms, then they will teach that monatomic molecules [*sic*] are not molecules. One colleague who piloted the materials in the companion volume on **Elements, compounds and mixtures** reported that the students 'have been taught that substances like Ne, Ar, etc have particles that are atoms (NOT molecules)'.

Physical and chemical change

The distinction between chemical and physical changes is often presented as important and unproblematic at secondary level (see Chapter 6). If this is so then science teachers should be able to agree on the classification of processes as chemical or physical and also on the reasons for their decisions (based upon the definitions they use). As part of the present project teachers were asked to undertake such an exercise (*ie* classifying changes as physical or chemical and explaining why).[32] As would be expected, many processes are not problematic in such a context. Freezing liquid nitrogen is agreed to be a physical change, and burning magnesium in oxygen is accepted to be a chemical change. However, such a common-place example as dissolving salt ('some sodium chloride is added to a beaker of water, and left to dissolve') did not lead to a consensus. This was commonly felt to be a physical process as;

- it 'can be reversed by physical means' and 'reversed easily';

- there were 'no structural changes to H_2O';

- there was 'no new substance made', 'no new compound', 'no change in chemical composition';

RS•C

- 'NaCl remains in ionic form only separated by H_2O molecules';

- one could 'extract NaCl out again unchanged'; and

- there were 'no apparent heat changes'.

Despite these views some teachers thought that this change could not be categorised:

'Although process is reversible, the sodium and chloride ions are separated when they are hydrated.'

'This is usually described as a physical change as it is easily reversed. However, according to some other ways of describing (*eg* in terms of bonding and types of substance produced) it could be put in the other category.'

One teacher wanted to count this process as both physical and chemical:

'BOTH! Solid NaCl easily recovered, but new particles formed (*ie* hydrated ions) which did not exist in original'

And some teachers feel this is a chemical change with 'bonds broken':

'I would argue (but not to my classes) that this is a chemical change – ionic bonds broken, ions hydrated.'

The definitions teachers were using provided criteria for making a judgement – easily reversible, new substances formed, bonds broken, energy changes – but these criteria were not adequate for agreeing on a definite classification (see Chapter 6). This is not an isolated example. Another process commonly observed in school laboratories (the heating of copper sulfate crystals to 'drive off' water of crystallisation) also divided teachers on whether it was a physical change, a chemical change or could not be categorised.

Implications for teaching chemical concepts

This chapter has tried to show why teaching chemical concepts is not a clear cut and straightforward matter. People have an aptitude for quickly learning new concepts such as 'tree' and 'animal', yet the process that helps us learn these 'natural categories' may not be very helpful when we need to learn scientific concepts that are usually delimited by technical definitions.[33]

When we do work with formal definitions (which we expect our students to understand, learn and apply) we often find that they are not that helpful for students. They may be incomplete and only have limited application. They may be wordy and require prerequisite understanding of other equally difficult concepts. Some definitions we find in the books students might refer to have both failings.

If full definitions are too involved for the students, and if partial definitions are only a stop-gap strategy, then a middle way is needed. Perhaps the best approach is to keep things simple, but to emphasise the provisional nature of the definitions used.

'The idea of oxidation is very important to scientists. When oxygen is added we say a substance has been oxidised. This is not the only way of deciding when something is oxidised, but this is useful for us now. At the moment you only need to know about this way of deciding if something is oxidised...'[34]

In this chapter I have also tried to show that even when we take such an approach there is a basic problem with the definitions which we teach and assess in our classes. Even when experienced teachers are asked to consider basic chemical concepts we find that these experts cannot agree on which definitions are accurate (let alone helpful to students). Presumably the teachers helping in this exercise had little trouble knowing when they were dealing with elements or molecules – but found definitions of these ideas problematic.

When teachers were asked to categorise familiar processes according to a distinction often taught at lower secondary level (*ie* chemical and physical changes) they used the accepted criteria from standard school definitions but came to different conclusions.

RS•C

I feel these exercises, generously undertaken by colleagues who sometimes had thirty years teaching experience, demonstrate a serious problem about teaching chemical (or other technical) concepts.

Definitions are very important in science, and have a role in learning: but we develop our scientific concepts by a more convoluted process. We only fully appreciate our definitions in the context of developing understanding of the very concepts they are supposed to help us learn.[35] Most definitions are of limited application, or obscurely worded, until we already understand them in their context through using the concept and its definition widely.

When teaching we must use definitions carefully. It may well be appropriate to introduce them at an early stage in a topic, to allow students to become familiar with them, but teaching a definition does not in itself teach the concept. The best that can be expected is rote-learning until the student has enough context to make the definition sensible.

If concepts cannot usually be effectively taught through definitions, then it becomes important to have a useful alternative approach to thinking about how concepts are developed. In the next chapter it will be suggested that scientific concepts cannot be meaningfully learnt in isolation, but need to be learnt in terms of their relationships with each other.

Notes and references for Chapter 2

1. Important types of schema are 'frames' (by which we model the structure of our physical environments such as rooms), 'scripts' (by which we make sense of routine activities such as eating a meal, shopping or teaching a class) and 'concepts'. See R. T. Kellog, *Cognitive Psychology*, London: SAGE Publications, 1997.

2. Clearly the child's brain is recognising trees according to some physical process. This could mean the brain is following rules that are too complex or subtle for us to have discovered – but this seems unlikely. Research into 'neural nets' show that fairly limited 'artificial brains' (using simple electronic components) can be 'trained' to recognise patterns – such as sonar signals reflecting from submarines – without having been programmed with any specific algorithm. This ability of nets of neurons (or transistors) only requires feedback on the success of previous attempts to make identifications. As the human brain is much more complex, and has structures already in place through evolution, this may explain how much successful learning of concepts occurs with very sparse feedback.

3. An extreme example is the learning of human languages. The paucity of the input available to a young child learning its mother-tongue convinced Noam Chomsky that the child's brain contains some structure which has evolved to readily acquire any language which fits a certain basic pattern (and therefore that all natural human languages have an underlying similarity despite obvious differences). This 'language acquisition device' is part of the cognitive apparatus that develops in all normal humans. The 'device' may not be a located at one specific site in the brain, but is still actually part of the physical structure of the brain. See S. Mithen, *The Prehistory of the Mind: a search for the origins of art, religion and science*, London: Phoenix, 1998.

4. This distinction may sound familiar to readers who have read about the work of Lev Vygotsky, who distinguished 'spontaneous' and 'scientific' (*ie* learned-through-teaching) concepts. See L. Vygotsky, *Thought and Language*, London: MIT Press, 1986.

5. D. M. Watts & J. Gilbert, Enigmas in school science: students' conceptions for scientifically associated words, *Research in Science and Technological Education*, 1983, **1** (2), 161–171.

6. Many of these terms were borrowed from everyday use, and then had a tighter meaning imposed upon them by physicists. This practice still continues, and while the use of terms such as 'charm', 'truth' and 'beauty' are not likely to lead to serious confusion between their scientific and everyday meanings, the idea of 'spin' (as in electron spin) is certainly a source of learning problems.

RS•C

7. C. Sutton, *Words, Science and Learning*, Buckingham: Open University Press, 1992.

8. K. S. Taber, Can Kelly's triads be used to elicit aspects of chemistry students' conceptual frameworks?, 1994 – available via Education-line, at **http://www.leeds.ac.uk/educol/** (accessed September 2005).

9. The reason for using three diagrams, and asking students to select an odd-one-out is to allow students to make discriminations when they are not sure of the appropriate labels for their ideas.

10. A. H. Johnstone & D. Selepeng, A language problem revisited, *Chemistry Education: Research and Practice in Europe*, 2001, 3 (1), 19–29, available at **http://www.uoi.gr/cerp/** or **http://www.rsc.org/Education/CERP/index.asp** (accessed September 2005).

11. K. S. Taber, Finding the optimum level of simplification: the case of teaching about heat and temperature, *Physics Education*, 2000, **35** (5), 320–325.

12. K. S. Taber, An analogy for discussing progression in learning chemistry, *School Science Review*, 1995, **76** (276), 91–95.

13. For example, in the National Curriculum for England, see *Science: The National Curriculum for England*, London: Department for Education and Employment / Qualifications and Curriculum Authority, 1999.

14. J. G. A. Raffan *et al*, *Chemistry for Modern Courses*, Sevenoaks: Hodder & Stoughton, 1975, 33.

15. Oxford Science Programme: *Materials and Models* Oxford: Oxford University Press, 1993.

16. Oxford Science Programme: *Materials and Models* Oxford: Oxford University Press, 1993.

17. P. Lafferty & J. Rowe, *Dictionary of Science*, London: Brockhampton Press, 1997.

18. J. G. A. Raffan *et al*, *Chemistry for Modern Courses*, Sevenoaks: Hodder & Stoughton, 1975, 27.

19. The science teachers were attending presentations about the RSC **Challenging Misconceptions in the Classroom** project which led to this publication, at meetings arranged by the ASE (Association for Science Education). The teachers completed a version of the classroom activity **Definitions** included in this publication, but with the inclusion of an additional item on 'monatomic molecule'.

20. D. W. A. Sharp, *The Penguin Dictionary of Chemistry*, Harmondsworth: Penguin, 1983.

21. K. Gadd & S. Gurr, *University of Bath Science 16–19: Chemistry*, Walton-on-Thames, Surrey: Thomas Nelson & Sons Ltd, 1994, 16.

22. E. B. Uvarov, D. R. Chapman & A. Isaacs, *The Penguin Dictionary of Science* (5th Edition), Harmondsworth: Penguin, 1979.

23. J. Morris, *GCSE Chemistry*, London: Collins Educational, 1991, 264.

24. D. W. A. Sharp, *The Penguin Dictionary of Chemistry*, Harmondsworth: Penguin, 1983.

25. The context for this definition was 'The smallest particle of matter which can exist in a free state (see atom). In the case of ionic substances such as sodium chloride, the molecule is considered as NaCl, which exists as an ion-pair in the gas phase, although the solid consists of an ordered arrangement of Na^+ and Cl^- ions.' See Chapter 7 for a consideration of ionic 'molecules'.

26. E. B. Uvarov, D. R. Chapman & A. Isaacs, *The Penguin Dictionary of Science* (5th Edition), Harmondsworth: Penguin, 1979.

27. P. Lafferty & J. Rowe, *Dictionary of Science*, London: Brockhampton Press, 1997.

28. E. B. Uvarov, D. R. Chapman & A. Isaacs, *The Penguin Dictionary of Science* (5th Edition), Harmondsworth: Penguin, 1979.

RS•C

29. The same is not true for free neutrons which have a surprising short half-life of a thousand seconds, *ie* less than twenty minutes: R. D. Harrison, *Nuffield Advanced Science Book of Data (Revised Edition)*, London: Longman: 1985.

30. Of course the stability of any species depends upon the conditions. Most molecular materials that are not gaseous will become so on heating, and so fit this definition. Continued heating (without the opportunity for reaction) will lead to gaseous atoms, but then if heating was continued the atoms would be ionised, and eventually the nuclei will break up, and even the protons and neutrons would not be stable at a high enough temperature. Under the conditions we normally consider, atoms do not usually have 'independent' existence.

31. This is not a new issue. A letter to the School Science Review in 1961 criticised an article defining the molecule as 'a group of atoms bonded to one another', and championed instead 'the smallest separate particle of a gas which moves about as a whole'. The letter's author pointed out that this definition of molecule would include 'species consisting of 1 atom only' – J. W. Davis, Molecules & Ions, *School Science Review*, 1961, **43** (149), 195–196.

32. Some responses were obtained from teachers attending presentations about the RSC **Challenging Misconceptions in the Classroom** project which led to this publication at ASE (Association for Science Education) meetings, and others by post from colleagues volunteering to trial materials.

33. W-M. Roth, Artificial neural networks and modelling, knowing and learning in science, *Journal of Research in Science Teaching*, 2000, **37** (1), 63–80.

34. More research is needed to find out how students will respond to such approaches, but based on what is known at present this would seem to be 'best advice'.

35. K. S. Taber, Time to be definitive?, *Education in Chemistry*, 1995, **32** (2), 56.

RS•C

RS•C

RS•C

3. The structure of chemical knowledge

This chapter considers the way knowledge is structured in chemistry, both formally and by individual learners. The chapter also includes a consideration of 'concept mapping', a simple and useful technique which can be used for planning lessons, for diagnosing aspects of how learners structure their knowledge, and as a useful study and revision tool.

Concepts take their meanings within knowledge structures

The previous chapter (on concepts in chemistry) discussed what is meant by a concept, and how such concepts may be formed so that learning can take place. One key aspect of looking at the meaning of a concept is an awareness that it is difficult to consider particular concepts in isolation. When we define concepts we usually do so in terms of other previously learnt concepts (see Chapter 2). It is clear, then, that the meaning of a single concept depends upon how we understand it in relation to other ideas. In other words, in order to understand what we mean by a chemical concept we need to see how it fits into a wider structure of ideas. We need to help students 'make the connections'[1] to develop their own conceptual frameworks. This chapter considers this idea of knowledge structures in chemistry.

Formal and personal ways of structuring chemical knowledge

One type of knowledge structure is that of chemistry itself. I will call this formal organisation of chemical knowledge the conceptual structure of the subject. This is the way chemical knowledge itself is organised 'officially'. We can find out about the conceptual structure of chemistry from journals and books. This type of structure is very important to teachers, both because in a sense it is the subject matter that we teach, but also because it is an important tool for planning effective lessons.

The second type of knowledge structure is that in the minds of learners: their own personal ways of relating and understanding chemical ideas. The way learners represent knowledge is very important. For one thing the students' existing knowledge structures are an important determinant of what and how they learn new material (see Chapter 4). Also, if we accept that concepts only take their full meaning in relation to each other, then we need to know about the way a learner makes relationships between different ideas before we can judge if they understand them as intended. The term cognitive structure is used to describe the way a person's knowledge is organised.

It is not possible to directly observe cognitive structure, so it must be inferred from indirect evidence (such as answers to the teacher's questions and the responses made in tests and other probes).

The conceptual structure of chemistry

A curriculum subject such as chemistry is not a set of isolated facts or principles. Chemical knowledge is structured. Not every chemistry teacher or chemist would agree on precisely what that structure is, and, anyway, it is a fluid structure. The structure of the subject now is different from 50, 100 or 150 years ago, as the subject has developed.[2]

RS•C

For example, the 'traditional' division of chemistry into inorganic, physical and organic branches (see Figure 3.1) has become less significant in recent years,[3,4] with much important work done across divisions (so we have physical organic chemistry as a field in its own right, and developments in areas such as organic conductors, and organometallic chemistry) and even across the boundaries of the discipline (in materials science, molecular science, biochemistry, geochemistry, *etc* – and a rather arbitrary distinction between some aspects of physical chemistry and the field of 'chemical physics').[5]

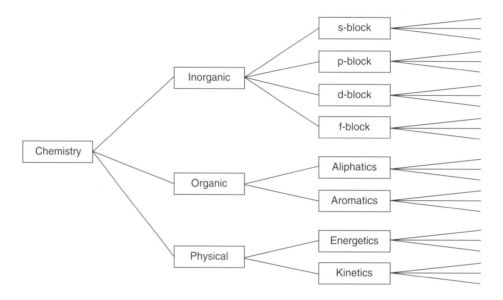

Figure 3.1 One way of structuring chemistry

Yet to some extent this break-down of the 'traditional' subject structure is merely a reflection of the advances in theory and the available techniques which have seen – for example – quantum-mechanical calculations and spectroscopic and related techniques become more important. Indeed, the development of the subject can largely be related to the increasing role of electronic structure as an organising theme for the subject.[6]

The structure of chemistry has shifted, but it no doubt exists. This is unavoidable, due to the nature of knowledge itself. As was considered in the previous chapter, all of our concepts take their meaning from the way they relate to one another (see Chapter 2).

To some extent we can see chemical concepts as hierarchical, starting with the most basic definitions and distinctions (some of which are considered in Chapter 6). So the notion of substance is fundamental – and allows us to define what we mean by chemical reactions (a theme discussed in more detail in Chapter 9) - and pure substances are divided into elements and compounds. The idea that matter is quantised (the molecular model) rather than continuous is another key tenet of the subject. Once we see matter in this particular way, a key concern becomes how the particles are arranged (*ie* chemical structure – see Chapter 7), and how they are held together (*ie* bonding – see Chapter 8) and how rearrangements may occur (*ie* reactions – see Chapter 9).

One might see other concepts as being at less significant levels of the hierarchy – so once the quantum model of matter is established, it is possible to look in more detail at atomic, and indeed nuclear, structure. Once we have a concept of element we can consider what we mean by metals and non-metal, and by finer distinctions such as transition metals or halogens.

However, we soon find that a hierarchical structure does not do justice to the sophistication of our subject! For example, our periodic classification ties the concept of element to ideas about atomic structure, and to the notion of substance. As well as dividing substances according to positions in the

RS•C

Periodic Table we have other classifications such as acids and bases, and oxidising and reducing agents (which may be elements or compounds).

It soon becomes clear that there is no simple hierarchical relation between our concepts, but rather that they are organised into a kind of web or net: a structure with many nodes (concepts) connected by a complex network of relationships. Needless to say, this very complexity makes teaching chemistry a complicated, demanding, but rewarding business.

Representing knowledge structures: the concept map

Textbooks, and lessons, inevitably present material in a linear fashion. A book has to place some material ahead of others, and a teacher has to introduce some ideas before others. The nature of our world – of space (on a page) and time (in a lesson) – does not provide any alternative!

Yet the nature of chemical knowledge does not readily fit into such a spatial or temporal sequence. A graphical approach – something more like a diagram – might be a more appropriate way of representing our knowledge about chemistry (and many other subjects). For example, synthetic reaction schemes are often presented in a graphical form.[7] A concept map is a useful graphical representation, that can be used for any information that does not readily fit a linear pattern.

There is a large literature on concept mapping[8,9,10,11] and related techniques, and there are many variations on how such 'maps' may be produced. Anyone who finds the ideas presented in this chapter particularly helpful may wish to read up further, but in my own practice I have found a simplistic approach sufficient.

Take a look at the concept map in Figure 3.2.

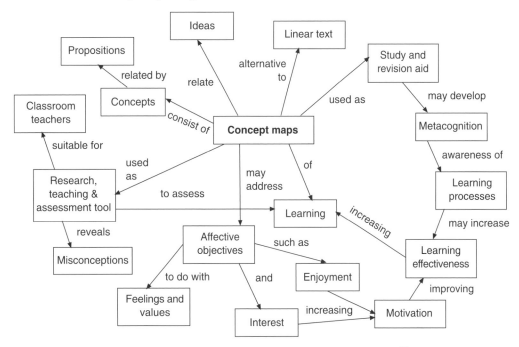

Figure 3.2 A concept map for the concept 'concept map'[12]

This is a concept map for the concept of 'concept map'. 'Read' the map. Note that there is no one and only correct 'order' for reading the information.

The map has two types of components:

■ nodes – representing 'concepts'; and

■ connections – representing propositions that relate the nodes.[13]

RS•C

Such a representation is a type of model – a graphical model of some aspect of human knowledge. As a model it is under the author's control (see Chapter 6). When preparing such a map you can (and indeed need to) make decisions about;

■ what the central concept will be;

■ the amount of information to be included;

■ the types of concept labels to include; and

■ the amount of detail required for connections; *etc.*

The central concept: in this case the concept is 'concept map', but it could have been 'chemistry' (and Figure 3.1 could be seen as a type of concept map), 'metals', 'oxidation', 'rates of reaction', *etc.* Do not be precious about what is meant by the word 'concept' (see Chapter 2) - if you can label an idea it probably counts as a concept.[14] It is useful to bear in mind, however, that such specific concepts are embedded in larger complex networks of ideas. In drawing a concept map we select the central idea and try to extract the most salient connections.

The amount of information: just like the party game of connecting two movie stars or rock musicians through a sequence of fellow artists they have worked with, or films/albums they have worked on, it is possible in principle to connect any concept with any other – by increasing the number of 'degrees of separation' allowed. Concept maps can become too complicated, and too dense with information to be useful, so it is important to be selective in what is included. (If a map is getting too dense it is always possible to replace it with several maps, which are inter-linked – just like city maps which have central sections reproduced in more detail at a different scale.)

Concept labels: it is best to keep the node labels concise, and familiar. Although concept maps avoid the need to arrange knowledge in hierarchies (like identification keys) some concepts have greater generality than others. For example, the concept 'base' is more general than the concept 'alkali'. Where a map includes concepts of different generality, the more general concepts are usually located more centrally, and the more specific concepts more peripherally. Some concepts are so specific that they are best seen as examples (*eg* 'sodium hydroxide'), and usually fit best near the edges of a concept map. Sometimes examples may be represented differently (say, not having a box around them) to show they are considered less fundamental to the structure being described.

Connections: a connection on a concept map is usually of the form of a proposition: that is a sentence which shows how the two concepts are connected. A line connecting 'alkali' and 'base' might represent 'an alkali is a type of base', or 'an alkali is a soluble base'. (Note that both of these propositions show that 'alkali' is a type of 'base' - but more specific information is given in the second example.) It is possible for the propositions to be written in full on the map, or abbreviated into note form, or just represented by a number or letter key – in which case the propositions may be listed separately. The propositions in Figure 3.2 are indicated by arrows, but often they will just be lines.

Concept maps are tools to help us represent (and explore and develop – see below) our knowledge, and should be used flexibly. It is not helpful to impose too many rules about how we should draw them.

Circular definitions or spiral curriculum?

In the previous chapter we saw that learning scientific concepts can be very problematic. To summarise some of the points made there:

■ chemical/scientific concepts are usually defined by rules;

■ the rules are often subtle and difficult to explain; and

■ defining the concept is usually only possible in terms of other concepts which also need to be defined.

RS•C

This can potentially lead us to a problem where we can only understand concepts in terms of the other concepts to be learnt – a 'vicious circle' (see Figure 3.3).

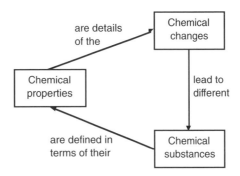

Figure 3.3 A concept 'circle'

So, for example we might define a chemical change as one which leads to new substances, but we would need to know how to distinguish substances to know if the product was different (or just the same substance in a different form). We can characterise a chemical substance in terms of its chemical properties, that is the details which chemical changes it undergoes. But this, of course, needs us to understand what a chemical change is... This 'concept circle' would seem to have no starting point.

There are many other examples that could be given, as the previous chapter suggests. Clearly, many of our students do acquire acceptable versions of such concepts, so in practice students can enter the 'concept circle'.

A key point to appreciate here is that learning chemical concepts is not an 'all-or-nothing' experience. We expect our lower secondary students to acquire a working concept of, say, chemical compound, but we appreciate that this understanding will at first be tentative, and even 'fragile'. As the student moves through the secondary school, and then perhaps through college and maybe even university study of the subject, we expect their appreciation of the concept to become, deeper and more robust.

As the student better understands one concept, he or she can begin to have a better appreciation for all the other concepts that are closely linked to it.

This is the rationale for having a 'spiral curriculum', where topics are introduced, and later revisited in increasing depth. This approach has two powerful advantages. Firstly we know that the actual process of 'fixing' concepts in the brain does not take place instantly. The permanent changes in brain structure which lead to long-term learning take place in the days, weeks and perhaps even months after the 'learning experience' in class. Revisiting a topic frequently may influence this process of 'laying down' permanent memory traces.

Conversely, ideas that are not used at all for months may not leave very strong traces in memory. Many good teachers use a 'drip-feed' approach: gently and briefly reiterating key ideas and concepts when a teaching opportunity arises. This reinforces learning both by emphasising the significance of the concept, and by providing links to other curriculum topics.

The second advantage of a spiral curriculum is that between successive stages of exploring a topic (such as acids) other topics will have been visited (perhaps combustion, metals, water) and new links can be developed which were not possible before. For example, if oxidation is studied first in the context of combustion, and subsequently the student studies displacement reactions between metals and metal salts, this provides a new context for revisiting and expanding the oxidation concept that was not available before.

Formal curriculum structures may include deliberate attempts to build upon the idea of a spiral curriculum, but will only be successful if there is genuine progression in the depth and breadth of the treatment each time a topic is studied. [15]

RS•C

Initiating a learning spiral

Seeing learning in terms of concept-spirals overcomes the obvious problem of concept-circles (where teaching any concept requires understanding of another, which in turn depends upon...). However, it is clear that it does not avoid the problem of how to get such a learning spiral going.

This means that the teacher needs a starting point that learners can relate to: something that is already familiar, and can be used as the substrate for new learning. (When such a connection is not made, useful learning is usually blocked – see the next chapter). The new material must be 'anchored' to this substrate by a suitable 'hook' – something that relates what the learner already knows with the chemical ideas to be acquired.

Sometimes there are obvious targets in the learners' existing experiences. To take an example: all students are familiar with fires (bonfires, burning candles, seeing house fires on the TV news *etc*), and so this makes a suitable starting point for introducing the chemical concept of combustion. The student can learn this label, and associate it with their everyday experience of burning. This can then later be a suitable hook for teaching about the concepts of 'chemical change', 'oxidation', and later 'exothermic reactions'.

Sometimes such obvious targets are not available. Often the obvious place to start is not within everyday experience, but in terms of formally taught prior learning (an approach which can go wrong when the prerequisite learning is not what the teacher assumes – see Chapter 4). On other occasions, when the new ideas are especially abstract or obscure to learners, it may be necessary to form links with existing knowledge by developing analogies from the familiar to the novel idea being taught. (The effective use of analogies needs careful planning, as is discussed in Chapters 7 and 10.)

Often the teacher needs to undertake a formal 'content analysis' of a topic.[16]

Concept maps as teaching tools: content analysis

Whenever a teacher has to teach a new topic it is sensible to undertake a form of content analysis, which:

- determines the precise ideas that need to be covered;

- how they are inter-related; and

- what other concepts will be needed to teach the new ideas.

The material to be covered is often laid down in curriculum documents or in a school or college's schemes of work. The teacher will need to decide in which order to introduce ideas, which is why a logical analysis of how the concepts are related is important.

However, it is just as important to consider the pre-requisite knowledge that is being assumed in teaching the topic. This is because:

(a) sometimes students do not hold the assumed prior knowledge (or hold alternative distorted versions of it), and so it is important for the teacher to check that students have an acceptable understanding of these ideas before setting out on the new exposition (many of the resources in this publication are probes suitable for diagnosing students' ideas).

(b) the links with prior knowledge need to be made explicit – sometimes these ideas are so familiar and obvious to the teacher that they are not emphasised enough for the students to spot them. This can lead to the new learning lacking 'anchors' in existing knowledge, and (to follow the nautical metaphor) floating away to become isolated icebergs of knowledge, or breaking up to give conceptual flotsam with no coherence or utility to the student, and conceptual jetsam washing up to form inappropriate links with other islands of knowledge. (These different types of learning blocks are discussed in the next chapter.)

(c) the new topic provides opportunities to reinforce many key ideas from previous teaching by the drip-feed approach (see above).

There are many ways to undertake such an analysis of content. One way is a kind of programmed learning approach, where the entire topic is reduced to a series of logical statements, and each of these propositions is sequenced so that each statement introducing a new idea is grounded in the earlier statements. If done effectively such an approach can be very useful, and it may well appeal to some teachers with particular 'thinking styles'. Some of us like to think through linear, logical steps – and may find such an approach useful and reassuring. (Do you like writing computer programmes, working on legislative committees, or undertake analytical philosophy for fun in your spare time? Some people do!)

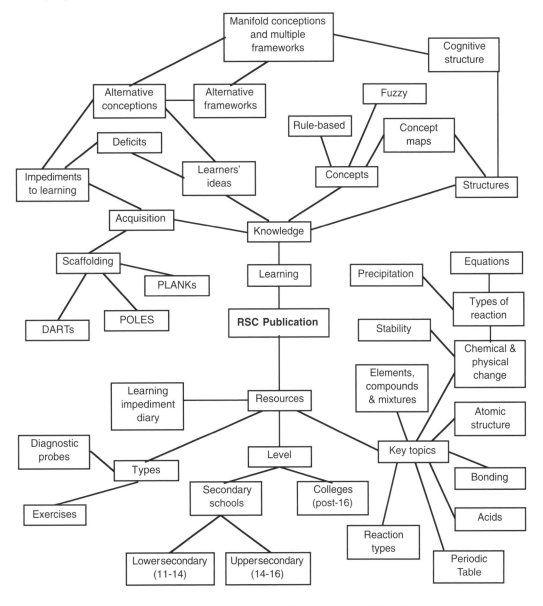

Figure 3.4 A way of representing the contents of this project

However, if such an approach does not appeal, you may find concept mapping as an easier way of working. Some teachers will prefer a visual way of representing information such as a concept map, rather than an ordered list of statements. For example, this publication includes a contents list, which is one way of finding out what it contains. However, the contents could also be shown in a diagrammatic form, like a type of concept map (see Figure 3.4).

RS•C

Some may prefer using a concept map, because they find constructing such a diagram much easier than just writing out sequences of sentences. Concept maps are a form of representation which lends itself to revision as understanding of a topic develops – something that has been found useful for teachers as well as students.[17] The concept map can be compiled in the order in which ideas hit you: start with your central theme, and just add things in until you feel you have covered the topic.

For example, consider the topic of acids as it might be presented at lower secondary level. The following concept map (see Figure 3.5, which is a reduced version of one of the classroom resource sheets included in this publication) could represent the teacher's plan for covering the topic. This map can then be a tool for planning individual lessons. (See, also, the concept map for hydrogen bonding in Chapter 5, Figure 5.3.)

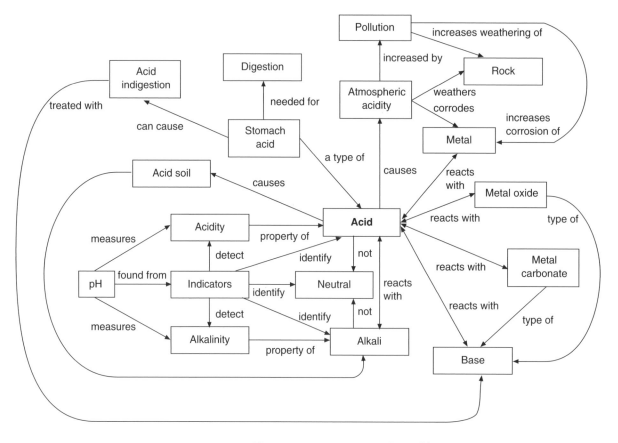

Figure 3.5 A concept map for 'acids'

Dimensions of cognitive structure

'Cognitive structure' is a way of labelling the ideas available to a learner. Using such a technical term might give the impression that this something that is well understood. In practice we do not really have a detailed understanding of how we store our knowledge.

One definition of cognitive structure is:

'the facts, concepts, propositions, theories, and raw perceptual data that the learner has available to her at any point in time, and the manner in which it is arranged'. [18,19,20,21]

A key point about this definition is that it includes the way knowledge is organised – as well as what knowledge is present. (If you have agreed with the earlier ideas in this and the previous chapter you will realise that this has to be so: the way concepts are related is a key part of what those concepts mean to us.) If we are interested in how well a student understands a subject like chemistry it is at

RS•C

least as important to know how they understand the concepts to be related as to know which concepts they have 'acquired'.

As has been pointed out above, concepts like oxidation or acid can be understood at increasing levels of sophistication as a learner passes through the school and beyond. A student could learn a definition of 'alkali' by rote, without necessarily having the understanding to apply the idea in appropriate contexts. Clearly questions such as 'has the student acquired the concept of acid' or 'does the student understand oxidation' are not very helpful unless prefixed by 'to what extent'.

Given that we are a long way from being able to 'read minds' directly and do not understand enough about how concepts are 'stored' within brains, we must rely on indirect evidence. Luckily, every time a student answers, or asks, a question, or writes about their chemistry, they provide such evidence. The activities in this resource are largely designed to be targeted at looking for evidence in key areas where we know students often do not understand topics the way we want. There has been a great deal of research to find out how students do make sense of scientific ideas (see Chapter 1) and that vast body of work provides a great deal of data about learners' ideas.

At one level this tells us a lot about which 'alternative conceptions' student commonly demonstrate. Unfortunately, however, different research, carried out by various researchers using disparate techniques, leads to different inferences about the way students store their ideas. The sensible (indeed common-sense) approach to interpreting this research is to take a view that learners' ideas vary along a number of dimensions, such as:

- tentatively held – deeply believed;

- alternative ('wrong') – conventional ('accurate');

- idiosyncratic – common;

- isolated conceptions – integrated frameworks; and

- unitary (consistent) – manifold (multiple frameworks).

We all have some ideas which are held quite provisionally (usually when the topic is not important to us, and/or the source is not considered reliable, and/or we are aware we only picked-up on part of the information: perhaps we hear the end of a radio news item about an election in some country we know little about). Other beliefs we treat as absolute matters of creed (eg England is joined to Scotland, and all people are entitled to be treated fairly in law). In the same way, students may hold some ideas quite tentatively ('I think manganese is a metal, but I'm not sure) and others very strongly ('I know reactions occur so that atoms can get full shells of electrons'). The strength of a conviction does not always relate to the accuracy of the idea!

Similarly, we each have topics where we know a few isolated facts, which are not strongly linked to anything else (how much do you know about romantic poets, or prohibition in the USA, or icons of the Russian church, or stone age hand tools, or Jung's notion of archetypes?) As science teachers, our knowledge of chemistry tends to be well integrated, coherent, and logically arranged, but many of our students have much more piecemeal knowledge (and they may find our knowledge of their favourite pop group/football team/television soap to be rather limited and fractured).

The final dimension above is not so obvious, but may be very important. People often hold alternative mental representations of the 'same' concept in their minds. It has long been known that some students seem to compartmentalise their formal school learning about some science topics separately from their everyday knowledge of the same topics (see Chapter 1). So a learner may show a good understanding of pH in a test, but refuse to drink something that is labelled acid in a normal social context, or may well know the difference between melting and dissolving in the laboratory, but talk of the sugar melting in the tea at home.

This phenomena of having separate scientific and 'life-word' versions of concepts has been seen as being related to the extent to which students integrate their ideas. Indeed it has been suggested that

RS•C

students' learning tends to be fragmentary and so their ideas exist as set of isolated 'minitheories'.[22] This, however, is unfair. Consider the following example of a (redrawn) student's concept map for the topic 'energy' (see Figure 3.6):

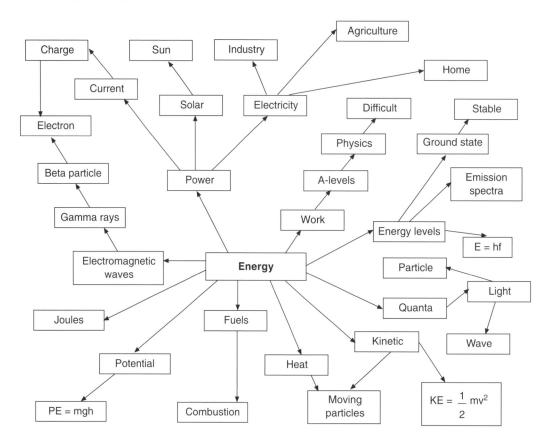

Figure 3.6 A student's concept map for energy[23]

Figure 3.6, whilst admittedly based on a (post-16) college student's concept map and not a younger student, certainly shows that students can link ideas effectively. We need to look elsewhere to explain why people should hold multiple versions of a concept. There are in fact good theoretical reasons to expect this.

For one thing, it seems that during evolution humans developed discrete 'domains' of knowledge related to the living world (thus the 'natural kinds' concepts for trees and animals discussed in Chapter 2); mechanics (thus the 'intuitive physics' in contradiction to Newton's laws) and social-psychology (enabling us to use our own feelings to model the feelings of others, and 'read' their minds from body language *etc*).[24] If this theory is correct, some compartmentalisation of our knowledge, with the possibility of forming multiple representations of some topics, is just 'human nature'.

Indeed, one explanation for how humans became so intelligent (compared with most other species) is that we developed the ability to take an idea, and mentally copy it, and adapt the copy to give a new idea, that is we developed the ability to think in terms of models and analogies![25]

This ability is essential to science. For example, all major scientific advances require someone to see ideas in a new way – for example, to see combustion in terms of reacting with oxygen rather than releasing phlogiston. Our internal 'conceptual construction kits' have an important advantage over mechanical construction kits. When scientists form a new theoretical approach they do not have to dismantle their previous understanding to use the same components in the new theory. Rather they build a new, separate, mental model, while still using the old one until the new one looks more promising (see Chapter 10).

RS•C

This ability is especially important in chemistry where we have many different models of major concepts like the atom, acids, oxidation *etc*. The scientist, and the learner, is able to use these multiple models because our brains are able to hold alternative versions of the same concepts concurrently.

The downside of this ability is that the learner does not always realise that she holds manifold conceptions or multiple frameworks, and so may not consciously select the most appropriate version to use in a particular context. A student who holds a scientifically sophisticated and 'correct' version of a concept will still sometimes produce an answer to a question using an alternative conception (see Chapter 8, Figure 8.20). 'Knowing' the 'right' answer does not imply knowing only the right answer. It is therefore important to test students' knowledge in a range of contexts.

Where we are trying to move students away from their alternative conceptions to more scientific ones, we may find that they 'relapse' in contexts that are especially familiar (where they habitually use their alternative conception) or where there is a high 'cognitive demand' due to the complexity of the problem (where a different aspect of the question may take their attention). This can be frustrating for teacher and student – and being told that such multiple representations may well be the basis of our intelligence may not seem reassuring to either!

Concept maps as diagnostic tools: eliciting knowledge structures

In view of this complexity that is characteristic of our knowledge structures, standard tests offer only a limited insight into student understanding. Even the probes in the companion volume are mostly focused on eliciting specific isolated conceptions (albeit ones which are common, and related to key topics). The best way to find out if a student really understands a topic is probably to spend extended periods of time interrogating them individually. Apart from being potentially intimidating for the student, this is an approach that teachers can seldom use due to time constraints.

A more time-efficient, and less stressful approach, is to ask students to prepare a concept map of their understanding of a topic. A concept map can reveal:

■ which key ideas are present/missing;

■ whether the student holds major alternative conceptions;

■ how well the student has integrated ideas within the topic; and

■ the extent to which the student links the concept with key ideas from related topics.

Introducing concept mapping

Before students can be asked to produce concepts maps they need to appreciate the idea of a concept map itself. Some students are likely to have met concept maps and related diagrams ('spider diagrams',[26] 'mind maps'[27]) before – but in some classes this will be a novel idea to at least some students. It is not appropriate to expect students to cope with the demands of a new way of representing information, and to think deeply about the topic to be mapped, so some familiarisation or practice will be needed.

There are various ways of approaching this:

1. the teacher might use and present her own concept maps (or examples from other classes, books *etc*) in teaching for quite some time before asking the students to produce their own;

2. students can initially be asked to produce practice maps on topics of choice (*eg* Manchester United; current fashion...) where the focus is on technique rather than content;

3. the teacher can lead class mapping of concepts with each student asked to suggest an addition to the map on the board/screen before asking students to produce group or individual maps;[28] and

4. the students can be provided with structured tasks using partially prepared concept maps, before producing them from scratch.

RS•C

This final approach, of 'scaffolding' (see Chapter 5) the task, can be done in a number of ways. Here we will briefly consider two examples that have been tested in schools.

A concept map for acids at lower secondary level

This task, **Revising acids**, was based around the concept map for acids produced above as Figure 3.5. The task was differentiated to place three levels of demand on the students. The most difficult level provided students with an outline map, with only the concepts shown (Figure 3.7):

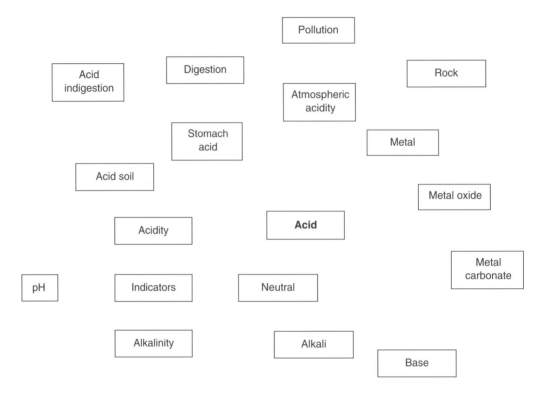

Figure 3.7 A partial concept map for 'acid'

The instructions for the students show them how to develop the map by adding connections.

When this exercise was carried out in schools to pilot the materials, it was found that some students were certainly able to make relevant and sensible suggestions for the logical connections between concepts (eg see Figure 3.8).

RS•C

RS•C

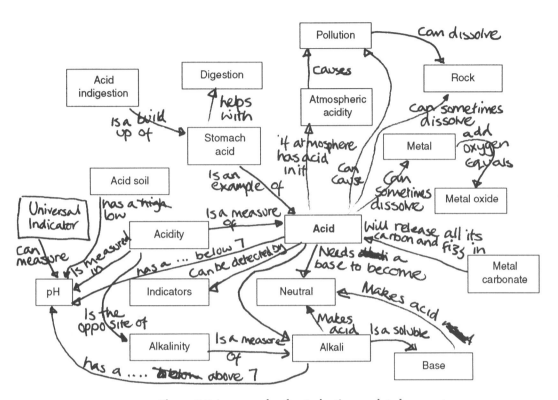

Figure 3.8 An example of a student's completed concept map

Although students may well miss some significant connections, and may use less precise or technical phrasing than the teacher would prefer (compare Figure 3.8 with Figure 3.5), they can also demonstrate considerable knowledge and understanding. (And, as after any learning activity, the teacher may provide students with a copy of a model answer – and this is likely to be more valuable once the students have made efforts to work through the ideas themselves.)

The example map shown above is from a student in a group of 12–13 year olds. This individual added a new concept - 'Universal Indicator'. Other students in the group added other concepts such as 'salts', 'acid rain', 'calcium carbonate' (connecting 'metal carbonate' to 'rock'), 'carbon dioxide', 'hydrogen', 'water', and 'phenolphthalein'.

As well as adding new concepts to the map, students may also suggest valid connections which were not expected. For example, among the suggested links from a student in a group of 13–14 year olds were:

■ digestion – pollution ('produces methane');

■ pollution – metal ('produces smoke when extracting'); and

■ metal carbonate – metal oxide – metal ('product of extraction').

Of course, student responses may also suggest they hold alternative conceptions. A student in the same group of 13–14 year olds proposed that an acid 'can be a base dissolved in water'

A less demanding version of the task provided an outline map with connections between the concepts (see Figure 3.9), and a separate list of incomplete statements representing the links (eg 'A. Acids in the _____ cause atmospheric acidity.') The students had to identify the links by labelling the connections on the map (as 'A' etc), and complete the statements.

RS·C

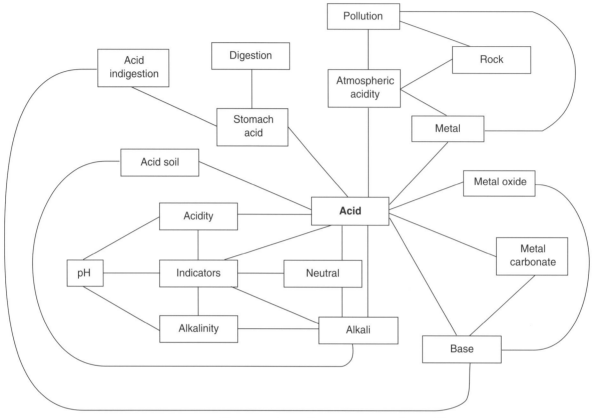

Figure 3.9 An outline concept map for 'acid'

A final, least demanding, version of the task provides the same map (see Figure 3.9) and the same list of statements, but this time complete (eg '1. Acidity is a property of acids'), so the student merely has to label the connections ('1' etc).

When the materials were used with classes some teachers felt that the least demanding version of the exercise was too simple, as it required no real knowledge of chemistry to complete. This was intentional, as it provides a task which only requires reading skills to complete, and so could be attempted with some success by a student who had learnt little about the topic (and could actually lead to some learning along the way – see the comments about DARTs, Directed Activities Related to Text, in Chapter 5).

The teachers setting this exercise were quite correct in their characterisation of this version of the task. However, the responses of some of the students asked to complete the intermediate version suggests that some teachers may have under-estimated the demand of the task (or overestimated the knowledge of the students) in deciding how to assign the three versions of the exercise. The following suggestions are some of the responses offered from 13–14 year olds in one class – the words in bold are those added by the student to complete the statement: [29]

■ Acids in the *metal* cause atmospheric acidity.

■ *Rock* can increase the rate of weathering of rocks.

■ Atmospheric acidity causes the corrosion of some *rocks*.

■ Too much stomach acid can cause *digestion*.

■ Too much stomach acid can cause *hydrogen*.

■ Some *pH* contain too much acid for many plants to grow.

RS•C

- Bases react with *alkali*.

- An *salt* is a base which dissolves in water.

- An *metal* is a base which dissolves in water.

- An *oxid* [sic] is a base which dissolves in water.

- Metal oxides are *acids*.

- Metal oxides react with *bases* to give salts and water.

- Metal oxides react with *alkalis* to give salts and water.[30]

- Metal oxides react with *metals* to give salts and water.

- Some *bases* react with acid to give a salt and hydrogen.

- Alkalinity is a property of *acids*.

- Neutral solutions can be identified using *acid*.

Perhaps some of these students would, in hindsight, have been better assigned to the least demanding version of the exercise.

A concept map for the Periodic Table at upper secondary level

Another concept map exercise, **Revising the Periodic Table**, prepared for this publication used a slightly different approach. This was a concept map on the topic of 'the Periodic Table' intended for upper secondary level students.

The students were provided with a concept map with labelled links (see Figure 3.10), and given a sheet to suggest sentences relating to the links. One example was completed for them ('6. An element is a single chemical substance') and they were asked to complete sentences for as many of the links as they could. This task is more demanding than two of the levels of the 'revising acids' exercise (which was intended for younger students), and differentiates by outcome, in terms of the sophistication of the responses students can make.

RS•C

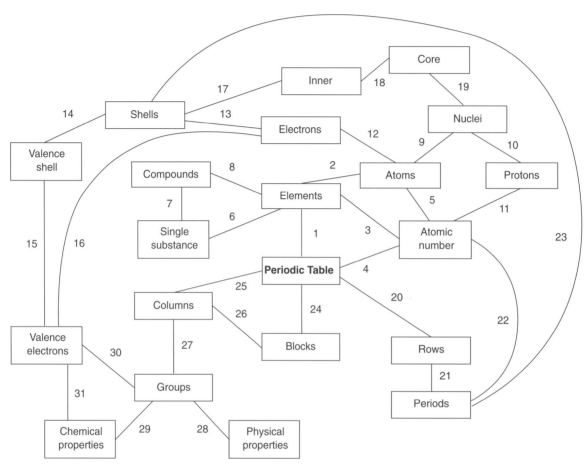

Figure 3.10 A concept map for 'the Periodic Table'

When this exercise was piloted for the project is was found that this format was helpful in allowing students to demonstrate their learning, and many sensible and valid responses were given. This type of exercise might be seen as providing a structure or 'scaffold' to help students explore and utilise their knowledge (see Chapter 5).

The format also provided examples of student statements which were unclear, ambiguous or wrong. The following statements are examples from two groups of 15–16 year olds:

■ 'The Periodic Table has many different blocks, each with an element in.'

■ 'The shell of electrons wants to be full. If not it is more reactive.'

■ 'A compound is not a single chemical substance.'

■ 'All elements are atoms.'

■ 'An element is a single substance, a compound is more than one.'

■ 'An atomic number shows the number of atoms.'

■ 'Atoms are another name for the same element.'

■ 'The amount of valence electrons depends on the chemical properties.'

■ 'The outer shell should contain 8 electrons to be full.'

■ 'There will be protons (the same number of neutrons) in an element.'

■ 'Protons go round the nuclei'.

■ 'The protons + electrons = atomic number'

These statements include the vague ('All elements are atoms'), and the sometimes-correct ('The outer shell should contain 8 electrons to be full' - only true for period 2) as well as the simply wrong ('A compound is not a single chemical substance'). Some statements may represent difficulty in expressing an idea. Perhaps 'the amount [*sic*] of valence [*ie* outer shell] electrons depends on the chemical properties' was meant to imply that the number of outer shell electrons determines the chemical properties. Maybe the statement that 'the protons + electrons = atomic number' was intended to imply that the number of protons equals the atomic number, and also that the number of electrons, equals the atomic number.

Whether such statements reveal genuine alternative conceptions, or limited language skills, or even just guessing in some cases, such responses reveal to the teacher areas where some attention is needed to check the student understands and can express the scientific idea.

Some alternative conceptions may derive, at least in part, from linguistic cues (the way the same or similar words are used in everyday life), or from some quirk in the way the a particular teaching scheme has been constructed. (These ideas are discussed in more detail in Chapter 4.)

However, often it is found that similar alternative conceptions are elicited from students in different education systems, even when the language of instruction is different. Marinella Spezziga, a teacher in Italy, translated the classroom materials discussed above, **Revising the Periodic Table**, into Italian, and presented them to her class of a similar age to the UK students. Marinella found some similar confusions and alternative conceptions among her students as have been elicited when the materials have been undertaken in English. As in the UK, some Italian students were unclear about the meaning of the terms substance and atomic number, the relative arrangement of sub-atomic particles, and the relationship between molar properties and atomic structure,[31]

- 'The combination of two or more elements is called substances'

- 'The atomic number is the number of atoms of an element'

- 'The atoms are found in the nucleus'

- 'Protons spin round nucleus'

- 'The electrons in the outermost shell have got the same properties'

There is a striking parallel between some of these suggestions, and those made by the UK students (reported above).

The materials discussed here, **Revising acids** and **Revising the Periodic Table**, illustrate two ways that concept mapping tasks may be set. There are other permutations on presenting incomplete concept maps. For example, a student could be given a list of propositions and asked to draw a concept map from them,[32,33] or given a nearly complete map with a few key concepts blanked out.

Concept maps as learning tools: learning styles

In many classes there will be some students who do not like, or even see the point, of concept mapping. Other students, however, are likely to find concept mapping activities enjoyable and useful.

We all have different preferred ways of learning (and teaching!), and just as some students feel they learn more by listening, and others by reading, some will prefer normal ('linear') textual materials, where others prefer more graphical (diagrammatic) ways of representing ideas.[34,35,36,37]

We need a lot more research about how important different 'teaching/learning/thinking styles' are in learning science, and what teachers should do to meet the needs of students with different ways of learning.

However, even without such research, there are a number of arguments for teachers incorporating techniques like concept mapping into their teaching. Given that it is likely that different students learn better from different approaches, then a teacher who uses concept mapping (along with flow charts and other forms of diagrams) as well as normal ('linear') text will:

RS•C

- be more likely to match up with different students' needs through the variety of approaches; and

- help familiarise learners with a range of approaches to help them find their preferred ways of learning.

In addition:

- different information is best represented in different formats; and

- multiple formats for the same information provide reinforcement and redundancy without obvious repetition. (See the comments on DARTs in Chapter 5.)

Finally, students learn best when they actively process information rather than passively receive it. Tasks that require students to translate information from linear texts to concept maps, or from flow charts to linear texts (for example) mean that the learner is actively involved in the task, is empowered to succeed (as the information is given, and just needs processing), and is working with the accepted version of the concepts (rather than just relying on existing understanding, which may include alternative conceptions).

An example of an activity which asks for such a 'conversion' is the exercise **Explaining chemical phenomena** (included in the section on **Scaffolding explanations** in the companion volume), where college students are asked to complete chemical explanations. Partial explanations are provided in flow chart form, which once completed need to be translated into linear text (see Chapter 5).

Concept maps as revision tools: metacognition

'Metacognition refers to a person's knowledge about his or her cognitive abilities[38] '

A final point about concept mapping is that it is a technique which can help students take responsibility for their own learning. Although the teacher may well feel that lower secondary students need a great deal of guidance, it seems appropriate to help students learn good study habits as early in their studies as possible. School sixthformers, college and undergraduate students are expected to take increasing responsibility for planning and executing their own study programmes, and those who leave school or college as effective self-directed learners will have an edge over those requiring 'spoon-feeding'.

This becomes especially important in terms of revision. Teachers can usually advise students on the progress of their work, but this is more difficult when students are revising, especially where the school or college gives study leave. Different students spending the same amount of time revising for a test may well use that time very differently. For those who tend to just read through their notes, there may be little awareness of how (in)effective their revision is until too late (*eg* the test). Where students can be encouraged to think about their own learning processes and develop an overview of their learning, they will have an advantage.[39]

The term 'metacognition' (being aware of one's own thinking processes – thinking about thinking)[40] may suggest a very high level skill, but can be practised in quite rudimentary ways. It is useful to encourage students to monitor their own progress and learning, rather than relying on external evaluation. Some revision activities are more likely to encourage such reflection than others. Students who are familiar with concept mapping may use it as a technique to test their knowledge of topics while revising, and may find it is a very helpful technique to help them judge their own progress: [41]

'My knowledge ... is very un-organised at present'

'I didn't realise how much the different areas inter-linked'

'I think this exercise was useful as it let me know exactly how much I know about [the topic], which I can now see is not enough'

Unlike some other activities, the non-linear nature of concept mapping allows the learner to move from working on one part of the map (if 'stuck') to others, and is an activity which inherently emphasises the structure of the subject:

RS•C

'Quite useful, brings back memories; good to see how well topics relate or how well you can interrelate them'

'I found I was digging around, trying to put fragments of things I could remember together. I found I could remember only scraps of information, but when doing the drawing [*ie* concept map], saw how things pieced together, and linked with other things'

'At first I did not know where to start but as I began putting ideas down, it reminded me of other points. I could have carried on writing'

'I didn't realise how much the different areas inter-linked. You could go on and on forever. I think this is a very pleasant experience and something I shall intend to continue doing.'

In any class there are likely to be some students who do not enjoy concept mapping activities (just as there are some who do not like producing written accounts, some who dislike calculations and others who prefer not to have to draw diagrams). Yet many students will enjoy and value the approach, and it certainly provides an alternative to formal note-taking. Moreover, it is a flexible technique – which can be used by teachers, individual students or groups; and as a tool for planning, for diagnosing understanding, as a learning activity or a revision technique. It is also a format closer to knowledge structures than the more usual text produced by teacher and student.

Notes and references for Chapter 3

1. B. Murphy, C. Murphy & B. J. Hathaway, *Basic Principles of Inorganic Chemistry: Making the Connections*, Cambridge: Royal Society of Chemistry, 1998.

2. H-J. Schmidt, Should chemistry lessons be more intellectually challenging?, *Chemical Education: Research and Practice in Europe*, 2000, **1** (1), 17–26, available at **http://www.uoi.gr/cerp/** or **http://www.rsc.org/Education/CERP/index.asp** (accessed September 2005).

3. One new approach to thinking about, and teaching, chemistry argues that 'the division of reaction chemistry into organic and inorganic is anachronistic and confusing' (note 4) and suggests that reactions be studied instead in the categories: redox chemistry; photochemistry; Lewis acid/base chemistry; radical and diradical chemistry. This approach is aimed at University level study, but is an interesting example of an attempt to reconceptualise the patterns of the subject. More information may be found at **http://www.meta-synthesis.com** (accessed September 2005).

4. M. R. Leach, *Lewis Acid/Base Reaction Chemistry*, Brighton: Meta-Synthesis.Com, 1999, 5.

5. For a recent survey of the frontiers of chemical science, see T. Lister, *Cutting Edge Chemistry*, London: The Royal Society of Chemistry, 2000.

6. W. B. Jensen, *Logic, History and the Teaching of Chemistry*, text of the Keynote Lectures, given at the 57th Annual Summer Conference of the New England Association of Chemistry Teachers, Sacred Heart University, Fairfield, Connecticut, 1995.

7. B. Earl & L. D. R. Wilford, *Further Advanced Chemistry*, London: John Murray Publishers, 2001, 201–207.

8. A. Al-Kunifed & J. H. Wandersee, One hundred references to concept mapping, *Journal of Research in Science Teaching*, 1990, **27** (10), 1069–1075.

9. J. D. Novak, Concept mapping: a useful tool for science education, *Journal of Research in Science Teaching*, 1990, **27** (10), 937–949.

10. J. H. Wandersee, Concept mapping and the cartography of cognition, *Journal of Research in Science Education*, 1990, **27** (10), 923–936.

11. B. Ross & H. Mumby, Concept mapping and misconceptions: a study of high-school students' understandings of acids and bases, *International Journal of Science Education*, 1991, **13** (1), 11–23.

RS•C

12. K. S. Taber, Student reaction on being introduced to concept mapping, *Physics Education*, 1994, **29** (5), 276–281.

13. A proposition can be defined as 'The smallest unit of knowledge that can stand as a separate assertion; that is, the smallest unit about which it makes sense to make the judgement true or false' – J. R. Anderson, *Cognitive Psychology and its Implications (4th Edition)*, New York: W. H. Freeman & Co, 1995, 145.

14. However, note that this does not imply that something is a concept only when you give it a name. Concepts may be 'tacit' - categories that we are aware of, and apply, at some level without being consciously aware that we are using them. All of our early concepts start of like this – we are able to make discriminations before we have any use of language, even though language, once acquired, becomes the key tool for learning, refining and exploring concepts.

15. In the English National Curriculum there are in-built opportunities to revisit many topics at successive stages. This has been criticised by some teachers who report that students find the curriculum repetitive. This should not be the case if each time a topic is met there is a brief recap of prior learning, and then significant progression in terms of the way the topic is developed. Teachers need to find different examples, and ways of presenting information, to ensure that students do not experience review as simply repetition. The QCA scheme of work *Science, A scheme of work for Key Stage 3*, QCA, 2000, may be useful in this respect, and the ideas about DARTs in Chapter 5 may also be helpful.

16. D. M. Gower, D. J. Daniels & G. Lloyd, Hierarchies among the concepts which underlie the mole, *School Science Review*, 1977, **59** (207), 285–299.

17. M. L. Starr & J. S. Krajcik, Concept maps as a heuristic for science curriculum development: towards improvement in process and product, *Journal of Research in Science Teaching*, 1990, **27** (10), 987–1000.

18. This definition from recent research (note 19) is an amalgam of ideas from two separate sources (notes 20 and 21).

19. K. S. Taber, Multiple frameworks?: Evidence of manifold conceptions in individual cognitive structure, *International Journal of Science Education*, 2000, **22** (4), 399–417.

20. D. P. Ausubel & F. G. Robinson *School Learning: An Introduction to Educational Psychology*, London: Holt International Edition, 1971.

21. R. T. White, Interview protocols and dimensions of cognitive structure, in L. H. T. West & A. L. Pines, *Cognitive Structure and Conceptual Change*, London: Academic Press, 1985, 51–59.

22. G. Claxton, Minitheories: a preliminary model for learning science, in P. J. Black & A. M. Lucas, *Children's Informal Ideas in Science*, London: Routledge, 1993, 45–61.

23. K. S. Taber, Student reaction on being introduced to concept mapping, *Physics Education*, 1994, **29** (5), 276–281.

24. It is deficits in this area which lead to autism.

25. A. Karmiloff-Smith, Précis of Beyond Modularity: A developmental perspective on cognitive science, *Behaviour and brain Sciences*, 1994, **17,** 693–745.

26. P. A. Wood, Spider diagrams, *School Science Review*, 1981, **62** (220), 550–553.

27. T. Buzan, *Use Your Head (Revised Edition)*, London: BBC Worldwide, 2000.

28. It is advisable to select a topic which everyone knows something about and feels they can contribute something. In introducing concept mapping with one class the central concept was 'blood' and 'blood is red' was accepted as a valid contribution. At this familiarisation stage it is the technique which is central, not the degree of insight produced.

29. National tests in England have shown that students at this age 'often fail to distinguish between the different substances with which [acids] react, metals, carbonates and alkalis/bases, and to identify the products of these reactions', Qualifications and Curriculum Authority (QCA), *Standards at Key Stage 3 Science*, London: QCA, 2001.

30. Students at this age are unlikely to have met the notion of amphoteric metal oxides which react with alkalis as well as acids.

31. Thanks are due to Marinella for re-translating the student responses back into English to share her findings.

32. T. Hudson, Developing pupils' skills, in M. Atlay *et al*, *Open Chemistry*, London: Hodder & Stoughton, 1992, 143–160.

33. W. Harlen, *The Teaching of Science*, London: David Fulton Publishers, 1992.

34. R. J. Sternberg, *Thinking Styles*, Cambridge: Cambridge University Press, 1997.

35. B. S. Thomson & J. R. Mascazine, Attending to learning styles in mathematics and science classrooms, 1997, at **http://www.ericdigests.org/2000-1/attending.html** or **http://www.stemworks.org/digests/dse97-4.html** (accessed September 2005).

36. H. Gardner, M. L. Kornhaber & W. K. Wake, *Intelligence: Multiple Perspectives*, Fort Worth: Harcourt Brace College Publishers, 1996.

37. R. Riding & S. Rayner, *Cognitive Styles and Learning Strategies: Understanding Style Differences in Learning and Behaviour*, London: David Fulton, 1998.

38. A. D. Pellegrini & D. F. Bjorklund, *Applied Child Study: a Developmental Approach (3rd Edition)*, Mahwah, NJ: Lawrence Erlbaum Associates, 1988.

39. C. Chin & D. E. Brown, Learning in science: a comparison of deep and surface approaches, *Journal of Research in Science Teaching*, 2000, **37** (2), 109–138.

40. Gunstone and Mitchell break metacognition down into metacognitive knowledge (about learning in general and about one's own personal learning characteristics), metacognitive awareness (of the purpose of, and progress in, a particular learning context) and metacognitive control (relating to the decisions and actions taken during a particular learning activity): R. F. Gunstone & I. J. Mitchell, Metacognition and conceptual change, in J. J. Mintzes, J. H. Wandersee & J. D. Novak, *Teaching Science for Understanding: A Human Constructivist View*, San Diego: Academic Press, 1988, 133–163.

41. The quotes are taken from K. S. Taber, Student reaction on being introduced to concept mapping, *Physics Education*, 1994, **29** (5), 276–281.

RS•C

RS•C

4.   Overcoming learning impediments

This chapter is concerned with the issue of looking at 'failures to learn', and in particular with developing a perspective and approach designed to see such 'failures' as an opportunity for the teacher to learn about how to help the learner.

The learning-doctor

A useful metaphor here might be to see part of the role of a teacher as being that of a learning-doctor. In other words – although it is disappointing when the desired learning has not taken place – the teacher's role here is to:

a) diagnose the particular cause of the failure-to-learn; and

b) use this information to prescribe appropriate action, designed to bring about the desired learning.

Two aspects of the teacher-as-learning-doctor comparison may be useful. Firstly, just like a medical doctor, the learning-doctor should use diagnostic tests as tools to guide action.

Secondly, just like medical doctors, teachers are 'professionals' in the genuine sense of the term. Like medical doctors, learning-doctors are in practice. (The 'clinic' is the classroom or teaching laboratory). Just as medical doctors find that many patients are not textbook cases, and do not respond to treatment in the way the books suggest, so many learners have idiosyncrasies that require 'individual treatment'. And just as General Practitioners in medicine may find interesting cases worth reporting to *The Lancet* or the *BMJ*, learning-doctors may well find interesting aspects of learners' ideas or their responses to teaching worthy of reporting to the profession in *Education in Chemistry* or the *School Science Review*.

The best science teaching, like medicine, is a research-based activity: both in terms of the teacher's craft being informed by published educational research, and in the sense that every new class (and every new learner) arriving to be taught a topic is a unique 'case' that needs to be approached as a problem to be solved using professional knowledge and skill.[1,2,3,4,5,6,7,8]

Obstacles to learning

There are many types of obstacle that can prevent a student learning as the teacher intends. Some but not all of these are – to a significant extent – within the teacher's control. Similarly, some of these obstacles are largely outside of the teacher's control, and of the focus of the present materials. However, it is useful to consider the various types of factor that can prevent students learning from teachers (if only to remind us how skilful successful teachers must be).

A hierarchy of obstacles

At a fairly basic level of analysis, we can identify the following reasons why a member of a class does not learn the material the teacher hoped would be learnt. These ideas owe a great deal to the psychologist Maslow,[9] but they are now familiar enough to count as common sense.

1. **Absence** If the student is not present, then no matter how wonderful the teaching, they will not learn anything. Although most of us do not suffer from Newton's attendance rates, (in some years no-one came to his annual lecture, on which occasions he felt justified in only lecturing for half the allotted time!), this is a common frustration for most teachers. Clearly there are ways of trying to help students who miss classes, but we would not generally hold the teacher responsible for those 'learners' who were not in class.

RS•C

2. **Physical conditions** If a student cannot see the board or demonstration, or cannot hear the teacher, then effective learning from the lesson is unlikely. The student may have a physical problem (such as needing spectacles or a hearing-aid), which the teacher may not be aware of. Bad hand-writing on the board or a class allowed to make too much noise would be the teacher's responsibility, but often conditions are outside the individual teacher's control (*eg* when the laboratory is not intended for such a large class, the teaching room assigned is next to a noisy drama studio, there is no screen available for the OHP, *etc*).

3. **Distractions** There may be more pressing issues in the learners' life than identifying the oxidising agent or calculating mole ratios. Clearly if a student is hungry, worried about a sick relative, scared of being bullied at break-time, apprehensive about the day's BCG inoculation, or in love with a classmate (or the teacher!) even the most skilful teaching display may not focus attention on the science. Often our teaching is not seen by class members as being the most important thing to think about, and in some circumstances this may even be a reasonable attitude for them to take.

4. **Motivation** As we all know, effective learning is only likely to take place if the students are motivated. Most students want to do well, want to feel good about their academic progress, and want to please teachers and family. Many are motivated to enter particular jobs or courses and are aware of the examination results they need. However, there are also many students in schools and colleges for whom there seems little reason to put in the effort to do well. And, sadly, there are some who are strongly motivated to be seen not to be valuing learning. Good teachers can sometimes get the best out of otherwise unmotivated students through the quality of their personal relationships with them. Similarly, by involving students in active learning (see Chapter 5), and presenting lessons in more interesting ways, much can be done to improve levels of motivation. However, sadly, there are some in our classes that are unlikely to be strongly motivated to learn from lessons even by the most gifted teachers.

Clearly all of the above are going to be substantive factors with some students in some classes, and these are not trivial issues. Indeed these factors are not entirely distinct, so, for example, improving motivation can reduce absenteeism. However the main purpose in outlining the problems above is to provide a demarcation between these issues and the main concern of the present chapter. Without wishing to underplay the importance of the problems discussed above (which will clearly be more significant in some institutions than others, and in some classes within institutions than others) the present chapter is mainly concerned with the reasons why students who attend classes, who are able to see and hear proceedings clearly, who are concentrating on the lesson and who are motivated to learn from it, should still often fail to do so.

This suggests that there must be at least one more type of obstacle to effective learning. It is helpful to label this as a communication problem, in the sense that the teacher's attempt to 'transfer' an understanding of some idea is thwarted.

5. **Not grasping the teacher's meaning** It is a common experience of teachers that an apparently logical, clear and coherent presentation of a topic, pitched at an appropriate level, to keen and capable students, who should have previously mastered any pre-requisite material, does not guarantee that the intended learning will take place. A whole gamut of evidence (such as homework, test responses, class questions) shows that communication often fails.

Even for the best teachers, the task of helping learners gain an acceptable understanding of some scientific ideas is often problematic. (Whilst this can be frustrating, it is also true that if the communication of concepts was a trivial process, then teaching would lose much of its challenge, and its potential for helping learners and so providing the teacher with job satisfaction.)

The remainder of this chapter is concerned with the nature of this 'communication breakdown', and how teachers should respond to such failures to achieve learning.

RS•C

Barriers to effective communication

It is helpful to begin our analysis of why science teachers' skilful expositions often fail to communicate the intended meaning to learners by considering an extreme case.

Imagine you have a new student in your class: a keen, intelligent student who has to date studied a curriculum comparable with the rest of the class, and who joins as you set out on teaching a new topic. Also consider, as sometimes happens, that this young person does not speak the language of instruction in your class, and that you do not speak the students' native tongue.

It would seem that there is little chance of even the most skilled teacher being able to effectively teach new science concepts to this student. Communication is about sharing understanding, and this is only possible if a common language can be found.

Even such a drastic case is not hopeless if at least one party is prepared to learn the language of the other (and I can recall one such case from my own teaching with a successful outcome due to the student's efforts with a translating dictionary, and a somewhat bilingual classmate acting as interpreter). The point is, that without a common framework for sharing meaning, a common language, effective communication will not occur.

The reason for presenting such an extreme case is to suggest that it stands as a suitable metaphor for all our inter-personal communication. It is what we – in science – might call the limiting case. Yet it is a potent metaphor for all our conversations with others: they are only successful to the extent that there is a common language for making sense of ideas.

By talking of a language I am not thinking so much of grammar, because most experienced teachers know how to keep their sentences simple enough for the age and ability of the classes they teach.[10] Although this may be a factor, far more important are the words we use. Not only which words, but what we mean by them.

Clearly science teachers know a great deal of technical vocabulary, and this has to be introduced sparingly and in a non-threatening way. But even with those words that are familiar to students, the meanings students have are often very different to those intended by the teacher.[11]

It is well recognised, for example, that technical terms used in science (acid, force, energy, momentum, plant, chemical...) often have much less rigid and tightly defined uses in common parlance.[12] Students also often have vague or inaccurate meanings for non-technical terms (such as omit, initial or abundant).[13]

What is perhaps given less thought, is how the meanings of a word vary from person to person according to all the manifold ways in which they experience it and use it. Indeed, there is a sense in which each individual English-speaking person speaks a subtly different language: as no two individuals share exactly the same set of word-meanings.[14]

What do *you* mean by a covalent bond?

Consider as an example a term like 'covalent bond'. Probably most students entering secondary school have no meaning for this term. As they pass through school, and possibly college chemistry and beyond, they construct a meaning, as they meet the term in a range of contexts.

Even ignoring students who get the 'wrong' meaning (perhaps mixing up covalent and ionic bonding) there is a whole spectrum of meanings to be developed. Perhaps initially 'covalent' bond might be understood as a pair of electrons 'shared' between two atoms (see Chapter 8), and this may originally be restricted to a few isolated examples (H–H, Cl–Cl, ...) until the concept is better mastered. It might be strongly associated with a line drawn between two chemical symbols, or a dot-and-cross type Lewis diagram. Perhaps it comes to have further meaning by being contrasted with the ionic bond.

The expert chemist, of course, brings a different and much richer meaning for the same term. For the teacher covalent bonds are at one end of a spectrum of bonds with varying degrees of polarity;

RS•C

perhaps they are associated with molecular orbitals formed by the linear combination of atomic orbitals; they are seen as bonding pairs which influence the shape of molecules in a slightly different way from 'lone' pairs of electrons, they imply something about the physical and chemical properties of the substances to which they are ascribed, *etc*.

A young student who has just learnt the notion of a covalent bond in a very limited context does not share the same set of meanings for the term as the teacher. This is not a case of the teacher being right and the student wrong, but of them having different concepts of covalent bond. The teacher and the student use the same word, but it is not clear that they refer to the same thing. The teacher's meaning is not only extended, it is more sophisticated, more subtle, and more deeply integrated into a framework of chemical ideas.

Now this situation is fairly obvious to teachers, and we all recognise that it is our responsibility to allow for the difference in meanings. The teacher tries to bear in mind the student's likely meanings for a word, and to hone his or her own language to both fit with, and ultimately to develop, the student's meanings. (This will be described in Chapter 5 in terms of seeing chemistry at the resolution available to the learner.)

The class from Babel

Now in any real teaching session, this difference in meaning is multiplied by the number of concepts being discussed. Every word the teacher uses referring to some idea is associated with a different range of meanings for each learner in the class. For each idea used in an explanation there may be thirty or so different understandings of what is meant, some quite close to what was intended, some less so. When this potential for understanding differently is taken over a whole class, over a whole lesson, it becomes clear why teachers have to become such effective communicators. Each classroom is a diluted version of the Tower of Babel. The student who does not speak the language of instruction is just the limiting case, in effect, every person in the class speaks a slightly different language.

How do we make sense of what people say to us?

Having described the problems of effective communication in the classroom, it seems appropriate to turn the discussion around, and consider how we ever manage to understand each other.

Each of us has a highly evolved and well developed tool for making sense of the world – our brains! The human brain (although it obviously has other functions) is a complex instrument for making sense of the world.

To a 'first approximation' it is useful to think about two different aspects of the learners' brain (although the distinction is certainly not an absolute one). One aspect of the brain that is clearly important is how it functions, what we may call the cognitive apparatus. Although our knowledge about how the brain functions to process information is far from complete, we know that different human brains tend to be generally rather similar in terms of how (for example) visual information is processed, or how memories are laid down and accessed. The processing of human language is also said to be very similar despite the apparent wide variation in human language. Brains obviously vary – whether through genetic predisposition, developmental maturity, or prior experience – but they seem to generally work the same way.

The second aspect of the brain that is important shows much greater variation. This is the individual's frameworks of understanding and knowledge, built up through a lifetime's wealth of experiences. Each eleven year old in a class has an enormous store of ideas, beliefs, images, memories, *etc*, that have been constructed in their brains through their personal life experiences. This complex framework of notions may be labelled as the learners' cognitive structure (see Chapter 3) and it is unique to that learner. Whenever a person listens to another, their ability to make sense of what is said will depend upon their unique cognitive structure, *ie* their existing frameworks of meaning.

RS•C

Constructing knowledge

One of the ways that brains tend to operate similarly, is that it is human nature to try and make sense of what is seen and heard. Indeed many common illusions depend upon the brain's ability to fit together a meaning from quite limited data. (So there may seem to be figures moving in the shadows; clouds may seem to take the shape of something familiar, such as a weasel {Shakespeare} or Ireland {Kate Bush}; and we readily recognise what quite minimal patterns (☺) are meant to represent).

Human memory is notoriously unreliable. Human memories are not accurate records of events experienced, but reconstructions. The brain is potentially swamped by vast quantities of data every second, yet actually only has the ability to consciously process a very limited amount of information at any time. The cognitive apparatus filters the vast majority of input before it reaches consciousness. The signals that do get through are not close to being raw data (except perhaps in cases of sudden pain!), rather they are already meanings that are imposed on the data to simplify it so the high level processing can cope. This is why we see figures moving in the shadows when no one is there. (In evolutionary terms, there is clearly an advantage to over-interpretation in this example, it is better to play safe and be alert.)

Consider two simple examples. What do you see below? (Figure 4.1)

Figure 4.1 Two recognisable patterns?

It is easier to describe what you interpret the two patterns to represent (a recognisable image and a familiar word) than to describe what you actually see. We find this type of effect in many aspects of life. Stereotypes are readily maintained as it is easy to find examples that seem to fit our prejudices. In science we soon learn to 'see' cells through a microscope, to 'see' isotopes on an NMR chart, or to see 'hysteresis' in a load-extension graph. It is human nature to develop more and more intricate conceptual frameworks to enable us to quickly make sense of increasingly complex phenomena.

Each of our students has constructed an extremely rich structure of conceptual frameworks through which he or she effortlessly interprets the world. This cognitive structure acts as the filter through which our teaching is heard. It is the substrate on which the learner builds a meaning for what the teacher has said. Often students construct meanings which are close enough to that intended for effective communication, but certainly not always. The teacher has to find ways to anchor new learning on the bedrock of the student's existing conceptual structure. To use a biochemical metaphor, the molecules of the teachers' message will only bind to the substrate (of existing conceptual frameworks) if they closely match the available target sites. If a binding site has the wrong structure (or is already occupied by an existing conception) then the intended synthesis cannot occur.

Talking in code

One analogy for the teaching process is that of communicating through code. If the communication takes place between two people who share the same codebook then the message can be successfully passed on. Human minds work through a form of electrical communication (mediated by chemical processes of course), yet communicate externally through signs and language (such as writing and talking). The brain has to re-code the electrical activity that is 'our thoughts' to be transmitted through speech or writing, to form a signal detected at another person's ears or eyes, where their brain will try and re-code the signal into the original meaning. Yet, despite strong similarities in cognitive apparatus, no one is born with the codebook in place![15] Each of us has to construct our own

RS•C

codebook by a process of trial and error; a process which is complicated by the fact that no two people we talk to are using exactly the same codebook as each other.

Luckily, it is often (but not always) clear when communication is not working, and in normal conversation we are usually allowed to have several attempts at making sense of each other until satisfied that a meaning has been communicated. The perceived social pressures of a classroom may however lead to less than optimum opportunities for this 'transactional calibration'.[16]

Often, however, the failure to communicate effectively may go unrecognised. If the listener does not re-construct the speaker's meaning, but still makes sense of what she hears, then neither speaker nor hearer will be aware that the message has been misconceived! Teaching can easily become an unintentional game of Chinese Whispers!

Learning impediments

This way of thinking about communication (or lack of it) during teaching suggests a way to analyse 'failures' to communicate. Such failures to communicate can be frustrating for teachers and students, whether they are clear at the time or only become apparent later. The following way of classifying learning impediments is intended to help the teacher decide how to respond effectively when such communication breakdowns are detected.[17] Sometimes this will help with the immediate problem detected with the current student or group, and sometimes this will be more useful in planning future teaching.

Successful communication occurs when the teacher's explanations are interpreted by the learner as having meaning sufficiently close to that intended by the teacher.[18] Apart from the more obvious barriers to this communication considered at the start of this chapter (the student is absent, not able to hear clearly, not paying attention *etc*), communication can also break down when the learners' 'coding apparatus' is sufficiently different from the teachers.

The teacher 'codes' his or her explanations from a background of chemical knowledge that is often much broader, deeper, more sophisticated and accurate from that of the students. However, the teacher uses her experience of teaching and of students to tailor the explanation to fit their current level of knowledge and understanding. Most of the time this is successful, but inevitably there are often occasions when at least some of the students 'decode' the explanation in ways that are not intended, or are unable to meaningfully make sense of the teacher's words at all.

In these situations we may think of the learning impediment being due to a lack of match between the actual knowledge and understanding of the learners, and that assumed by the teacher. This 'failure to match' can occur in different forms, and the teacher's next step depends on the particular type of mis-match.

The basic distinction is between the student failing to make any sense of the teacher's words, and in misinterpreting them.

Null learning impediments – causes of not understanding

A null learning impediment describes the situation where meaningful learning does not take place because the learner does not make a connection between the presented material and existing knowledge. In this case the teacher is assuming that the explanation will be interpreted in the light of some existing knowledge and understanding, but this does not happen, and the teacher's words do not make sense to the student. The learner does not make the intended connection.

Substantive learning impediments – causes of misunderstanding

The second type of situation is where the learner does make a connection with existing knowledge and learning, but not a useful connection from the point of view of the teacher. This usually means that the learner holds some alternative conceptions of the topic area, and understands the teacher's words in this inappropriate context.

RS•C

It is important to distinguish between these two types of problem, because the teachers' appropriate response is different in the two situations. In one case, new information needs to be added to the learner's existing knowledge base. In the other case some existing ideas needs to be challenged or developed (see Chapter 10).

Moreover, each of these two main types of learning impediment can be further sub-divided.

Two types of null learning impediment

Learners fail to make sense of the teacher's exposition because they have not been able to connect the teacher's words with their own existing knowledge. This could mean the learner does not have the prior knowledge, or that he or she just fails to realise what is being talked about! (So if a new teacher gives an explanation in terms of 'the valency shell', the learner may not realise this what the previous teacher referred to as 'the outermost shell'.)

Deficiency learning impediment Sometimes learners will not have the assumed prior knowledge. They may have been absent for some reason, or perhaps a previous teacher/school did not cover the material. (Or they may have made no sense of the teaching on an earlier occasion so that no significant learning occurred.) In this situation the appropriate response is to provide some form of suitable remedial input so that the learner acquires the 'missing' learning.

Fragmentation learning impediment However, it may be that relevant material is held in cognitive structure, but that the learner does not appreciate its relevance, so the new material is treated as an unrelated fragment of knowledge. I will describe this case as a fragmentation learning impediment. The most appropriate response from the teacher is to work to make the connection. This may simply mean asking the learner about the assumed prior knowledge and explicitly showing how the new ideas fit. Sometimes a more creative approach may be needed, with the teacher using analogies, metaphors and models to show that the new information is just like something already familiar to the learner. (Although important, this approach can lead to new alternative conceptions unless carefully planned. The example of modelling the atom as a tiny solar system is discussed in Chapter 7.)

Two types of substantive learning impediment

Learners come to classes with all manner of alternative conceptions, deriving from various sources,[19] and so there is great scope for new teaching being misconstrued in terms of existing knowledge and understanding. Substantive learning impediments are more serious than null learning impediments for two reasons:

(a) it is easier to 'fill' a 'gap' in knowledge than to challenge and replace an existing conception (see Chapter 10);

(b) 'gaps' in knowledge are often easily detected as learners and teachers can readily spot that no meaningful communication has occurred. Misconceptions may go undisclosed for long periods as both parties believe they understand the matter in hand.

In terms of helping individual learners it is important to identify and then challenge alternative conceptions,[20] and it is not that significant how the alternative ideas developed.

However, taking a more long-term view, it is useful to identify when alternative conceptions have developed from previous teaching. This makes little difference to those learners misunderstanding this year's lessons – but it may be possible to avoid the problems recurring with future classes. To revisit my notion of the learning-doctor, the medical doctor's immediate task is to diagnose and treat the patient's problem – but individual cases may also provide more generic information to improve public health. The slogan 'prevention is better than cure' can apply to teaching as well as to medicine.

Intuitive science conceptions – ontological learning impediments This is a (rather awkward) term for those alternative conceptions that arise from the learners' experiences of the world.[21,22,23] Perhaps the

RS•C

best example (because it has been found to be so widespread) is the naive physics conception that objects stop moving unless constantly pushed. This is an understandable deduction from everyday experience (as it is actually what happens in practice!), and causes many students difficulty when they study Newton's laws in school.

As students are not taught Newtonian mechanics until after they have experience of pushing objects around in a gravity-rich and friction-rich environment, it is inevitable that many will come to school science holding an 'impetus' framework (*ie* that when pushed objects move so far until the 'push/force/...' is used up). Physics teachers just have to accept this, be aware of it, and deal with it!

Mis-learnt science conceptions – pedagogic learning impediments There are doubtless many such alternative conceptions that arise outside of school, and which teachers can only tackle after they have been acquired. However, it has been suggested that in chemistry many of the alternative conceptions learners hold are 'pedagogic learning impediments' that derive largely from the teaching they have received.[24,25] A learner's personal beliefs about force and motion may be due to early life experiences, but it is much harder to explain why a learner would come to school believing that the sodium chloride lattice is comprised of diatomic molecules. Such ideas clearly develop from the way the subject is taught (see Chapter 10).

Sometimes these ideas are the result of students working beyond their level. Keen students may read ahead and can misinterpret material for which they have inadequate background knowledge. (Consider, for an example, what a typical 13 year old recently introduced to a basic model of atomic structure, might make of a laboratory poster showing the shapes of atomic orbitals.) Often, however, the teaching may not take the students' existing ideas into account sufficiently. If learners lack the expected prerequisite knowledge (see Chapter 3), or if the complexity of presented material overloads working memory, or contains logical steps that are too large for the learners to construct the teacher's meaning (see Chapter 5), then the learnt version of the ideas will not match what is intended.

If learners have alternative ideas of this type, the teacher needs to address them in the same way as intuitive science conceptions. However, it may be possible by re-thinking teaching (the order of topics, the emphasis given to certain ideas, the stage at which formal definitions are introduced, *etc*) to reduce the extent to which these problems are found in future year groups. Knowledge of how this years' learners have misunderstood concepts can be very useful in planning how to introduce those topics in future years.

The typology of learning impediments described above is represented in Figure 4.2. These characteristics of the types of learning impediment, and the actions indicated, are summarised in Table 4.1.

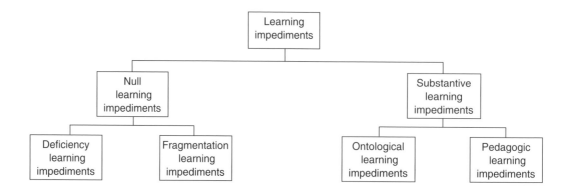

Figure 4.2 Types of learning impediment

RS•C

Type of learning impediment	Nature of impediment	Action required
Deficiency impediment (missing knowledge)	No relevant prior knowledge and understanding	Remedial teaching of prerequisite learning (if available), or restructuring of material with bridging analogies *etc.*
Fragmentation impediment (disconnected knowledge)	Learner does not see relevance of existing knowledge to presented material	Teacher should make connections between existing knowledge and new material explicit
Ontological impediment (intuitive science)	Presented material inconsistent with intuitive ideas about the world	Make learner's ideas explicit, and challenge them where appropriate
Pedagogic impediment (mis-learnt science)	Presented material inconsistent with ideas deriving from prior teaching	For individual learner: treat as ontological impediment; for future: re-think teaching of topic – order of presentation of ideas, manner of presentation, *etc.*

Table 4.1 Types of learning impediments

Applying ideas about learning impediments in the classroom

The section above discussed how failures to learn may sometimes be seen as breakdowns in communication due to a mis-match between the ideas the teacher expects the students to bring to class, and their actual knowledge and understanding.

The purpose of discussing such ideas, and in particular of suggesting a way of classifying different types of learning impediment, was to provide a way of thinking about learning difficulties that may be a useful tool for teachers.

Common and not-so-common alternative conceptions

Give a teacher a probe and you help him or her identify specific conceptions; teach a teacher to be sensitive to students' conceptions and you provide insight for life. [26]

The probes and exercises that are included in the companion volume have been written to take account of alternative conceptions that have been uncovered in research. There are some common alternative conceptions that research suggests are found in most classes in most schools and colleges. The materials have been prepared to help teachers diagnose and challenge some of these common conceptions. The two criteria that have been used to select topics for probes are:

(a) the topic seems to be commonly misunderstood in ways that can be readily identified; and

(b) the subject matter is significant for the understanding of basic concepts.

It has not been possible to deal with all of the alternative conceptions reported in the literature. Just as important, every learner is unique, with his or her own individual network of ideas, beliefs *etc.* So many learners have alternative conceptions that are idiosyncratic, and which cannot be revealed by the use of standard sets of diagnostic tools. So while it is hoped that the materials included in this resource will be useful, they will not provide a universal panacea for identifying students' alternative conceptions.

RS•C

The most important diagnostic tool: the teacher's sensitivity

In practice all teachers regularly spot learners' alternative conceptions. Often 'different understandings' are apparent in test responses or homework assignments. It is obviously more useful if the teacher can identify learners' alternative conceptions as early as possible. It would be ideal to have diagnosed and catalogued all relevant alternative conceptions (as well as having checked that pre-requisite prior knowledge is in place) before starting a topic. In practice this degree of auditing prior learning is not usually possible, although techniques such as concept mapping (as discussed in Chapter 3) can be very useful.

However, a good teacher can use classroom questioning to elicit many potential 'failures of communication' in situ, which allows the misunderstandings to be dealt with immediately, rather than when reviewing written work (by which time fanciful interpretations will have been rehearsed and may have taken hold in the learner's thinking).

The teacher's sensitivity to learners' potentially unhelpful ideas about science topics may be increased in a number of ways:

(a) with increased teaching experience there are more opportunities to be familiar with the types of ideas students use in their work;

(b) being more aware of the types of ideas that have been found and reported in research;[27]

(c) taking time to sit down with individual learners or small groups and exploring their ideas in a non-threatening context;

(d) developing a teaching approach that encourages learners to discuss and critique their ideas; and

(e) developing classroom questioning techniques which explore learners' interpretations in more depth, rather than simply evaluating responses as correct or not.

In particular the teacher has to try and interpret the learner's comments in terms of his or her own meanings, and not assume that the learner means much the same as the teacher hoped, or see apparently non-sensible suggestions as necessarily confused or meaningless. (Of course students do often answer questions with a 'random' or confused response, but some comments that seem meaningless may indicate that the student has an alternative conception for the point being discussed.)

The typology of learning impediments discussed above (see Figure 4.2) is meant to be a tool to help teachers think about learners' apparent failures to understand our teaching. The classification is not meant to be absolute, but to provide the teacher with a simple analytical framework. It is intended that using the framework will help develop sensitivity.

Included in the companion volume is a resource to help teachers work through this process, the **Learning impediment diary**, but the intention is to increase awareness of, and sensitivity to, learners' ideas, rather than to learn to use the typology itself.

The basic format of this 'exercise' is to keep a diary of the 'failures of communication' that you notice in your teaching, and to then try and classify these (and so start to think about their origins, and how they can be overcome and perhaps avoided in future). Some readers may feel that their teaching experience and sensitivity is such that this will not be a useful exercise: but it is offered for those who may find it helpful.

Keeping a diary of learning impediments

The basic form of the exercise is to keep a record of the learning impediments that you notice in your teaching. It is not necessary that you spot and record every occasion a learner does not understand the work. Indeed it may initially make more sense to decide to look out for one instance in each lesson, or one example per day. It is the analytical process that is important, not the quantity of examples you can spot. If you find this exercise helpful, you may decide to continue the diary

RS•C

indefinitely. Or you may feel you have become sufficiently sensitised to learners' ideas to be able to respond flexibly without needing to continue to use the diary. Or you may feel you only need to use the diary when first meeting a new class, or teaching a topic you have not met for some time.

1. Spot a failure to communicate

The first step is to be aware of occasions when a learner has not followed your intended meaning. This is easy if the student looks blank and is waving her arm to tell you she 'does not get it'. When students are (sadly) less concerned to understand, or are embarrassed to be seen to not 'get it', or when they misunderstand, then active questioning is needed.

The questioning is usually better if conceptual rather than factual ('can you explain why?' rather than 'do you know what?'), and an initial suggestion of a misconception may need to be probed (gently) by a short sequence of questions.

2. Detail the failure

The pressures of the classroom make it tempting to respond to any apparent misunderstandings by quickly providing the 'right' answers. However, without exploring the reasons for the misunderstanding such an input is often like 'papering over cracks' and will not correct the problem in the long term. A detailed exploration of what the students thinks, and why, will reveal more about how communication has failed, and, therefore, how to best deal with it.

3. Apply the framework

The simple 'key-type' flowchart (Figure 4.3) will help you work through what you need to know in order to respond effectively. In practice you will want to deal with problems as they arise in class, and probably will not have time to document examples at the time. However, it is suggested that it may be useful for you to use the flow chart as an aide-mémoire, and to complete the diary entries as soon as possible after the class.

You may find that it is difficult to analyse and record classroom instances of 'communication breakdowns' because of the pressures of working with large, demanding classes. An alternative approach (at least, as a starting point) would be to identify apparent problems from missing/wrong answers in students' work, and then ask to speak to the individuals about the work for a few minutes after class. This would enable the problem to be analysed in detail in a calmer environment, with less potential for embarrassing the learners, or of loosing 'the thread' of the lesson for the rest of the group.

RS•C

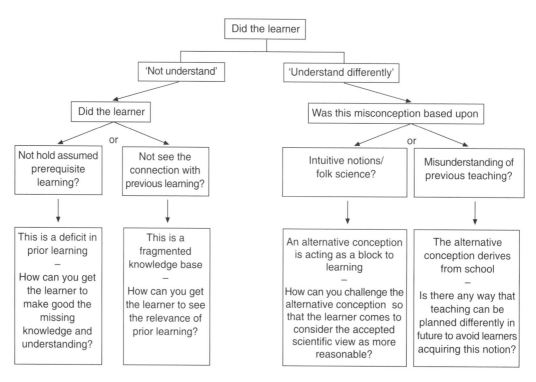

Figure 4.3 A flowchart for analysing learning impediments

Notes and references for Chapter 4

1. Educational research has been criticised for often being irrelevant to practice – a charge that has been strongly resisted by the research community (for example, the British Educational Research Association – see note 2). Within science education, research has been seen as an important way of informing practice for many years (*eg* notes 3 and 4). A recent book discusses a number of areas where research in science education has significant implications for practice (see note 5). The UK Association for Science Education (ASE) has been concerned about the issue of whether science teachers should be more involved in, and aware of, research for some years (*eg* see note 6), and there have been calls for science teachers to see their work as research based (see note 7). The RSC **Challenging Misconceptions in the Classroom** project which led to these present publications was very much seen as an attempt to help teachers adopt research findings in their classroom practice (see note 8).

2. P. Mortimer, letter to the editor, *Sunday Telegraph*, 28.11.99, reprinted in *Research Intelligence*, 1999, 70, back cover.

3. P. F. W. Preece, Towards a science of science teaching, *School Science Review*, 1977, **58** (205), 801–806.

4. P. F. W. Preece, A decade of research in science education, *School Science Review*, 1988, **69** (248), 579–586.

5. R. Millar, J. Leach & J. Osborne, *Improving Science Education: the contribution of research*, Buckingham: Open University Press, 2000.

6. K. S. Taber, Communication, motivation, ownership and subversion?: Facilitating the application of research in science education, 1996, available via Education-line, at **http://www.leeds.ac.uk/educol/** (accessed September 2005).

7. K. S. Taber, Should physics teaching be a research-based activity?, *Physics Education*, 2000, **35** (3), 163–168.

RS•C

8. K. S. Taber, Constructing chemical concepts in the classroom?: using research to inform practice, *Chemistry Education: Research and Practice in Europe*, 2001, **2** (1), 43–51, available at **http://www.uoi.gr/cerp/** and **http://www.rsc.org/Education/CERP/index.asp** (accessed September 2005).

9. D. Child, *Psychology and the Teacher (4th Edition)*, London: Cassell, 1986.

10. Techniques for estimating the reading age of classroom texts and teaching materials suggest that in science lessons reading tasks are often set at too high a level. However, it is not clear how definitive such findings are. Some techniques used give words with multiple syllables a heavy weighting as they are assumed to be inherently difficult. Science tends to use a lot of long technical words, and so texts will tend to have a high reading age on such measures. Although long words will tend to be more difficult to read, this may be less of a problem once those words are a familiar part of a students' vocabulary – see the comments in Chapter 5 on 'chunking' information. More significantly, perhaps, measures of reading age assume that such text is intended for lone reading by an individual. If a learner is reading along with peers or a teacher then it may be appropriate for the reading age of a text to be set above the individual's own reading age. (In terms of 'scaffolding learning', discussed in Chapter 5, the reading age of the text should be within the learner's 'zone of proximal development', to provide an achievable challenge that will help the learner develop their reading skills in science.)

11. D. M. Watts & J. Gilbert, Enigmas in school science: students' conceptions for scientifically associated words, *Research in Science and Technological Education*, 1983, **1** (2), 161–171.

12. C. Sutton, Language and communication in science lessons, Chapter 4 in C. R. Sutton, & J. T. Haysom, *The Art of the Science Teacher*, Maidenhead: McGraw-Hill, 1974, 41–53.

13. A. H. Johnstone & D. Selepeng, A language problem revisited, *Chemistry Education: Research and Practice in Europe*, 2001, **2** (1), 19–29.

14. C. Sutton, Science, language and meaning, *School Science Review*, 1980, **62** (218), 47–56.

15. Although it is conjectured that some brain structure acts as a 'language acquisition device', this is meant to channel human language within a general form: it does not provide the specific grammar (or lexicon!) of particular languages.

16. The idea that in dialogue we provide each other with 'constant transactional calibration' has been discussed by Jerome Bruner. See Chapter 4 of J. Bruner & H. Haste, *Making sense: the child's construction of the world*, London: Routledge, 1987.

17. K. S. Taber, The mismatch between assumed prior knowledge and the learner's conceptions: a typology of learning impediments, *Educational Studies*, 2001, **27** (2), 159–171.

18. Sufficiently close, that is, to be able to give answers that are judged correct in test, examinations, *etc.*

19. One classification refers to preconceived notions; non-scientific beliefs; conceptual misunderstandings; vernacular misconceptions; and factual misconceptions – *Science Teaching Reconsidered: misconceptions as barriers to understanding science*, at **http://bob.nap.edu/readingroom/books/str/4.html** (accessed September 2005).

20. Although it may be necessary to challenge students' ideas, research suggests that well established ideas are not readily 'forgotten', even once they are shown to be inadequate. See Chapter 10.

21. 'Ontological' being to do with the things that exist in the world and how they are categorised. For example, in some cultures, spirits and ghosts may be seen to be as real as living people (see note 22). In science education it has been suggested that many alternative conceptions are based on students' ontologies having category errors – such as classing heat as a substance like water (see note 23).

RS•C

22. K. S. Rosengren, C. R. Johnson & P. L. Harris, *Imagining the Impossible: Magical, scientific and religious thinking in children*, Cambridge: Cambridge University Press, 2000.

23. M. T. H. Chi, J. D. Slotta, & N. de Leeuw, From things to processes; a theory of conceptual change for learning science concepts, *Learning and Instruction*, 1994, **4**, 27–43.

24. M. Carr, Model confusion in chemistry, *Research in Science Education*, 1984, **14**, 97–103.

25. K. S. Taber, Prior learning as an epistemological block?: The octet rule – an example from science education, 1995, available via Education-line, at **http://www.leeds.ac.uk/educol/** (accessed September 2005).

26. With apologies to a well known charity slogan.

27. In England there is a government funded initiative to raise standards among the 11–14 age range. This 'Key Stage 3' strategy for science is intended to provide opportunities for 'continued professional development' to all teachers working with that age range in maintained schools. One of the units that is being developed is on 'key ideas and misconceptions'. There are strands in literacy, numeracy, science, ICT and critical thinking.

RS•C

RS•C

5. Scaffolding learning in chemistry

This chapter is concerned with the issue of how much help teachers should give learners *ie* the balance between 'spoon-feeding' and expecting students to cope without support. The notion of 'scaffolding' learning is discussed – that is providing support that is gradually reduced as the learner masters the material – and the technique of providing learners with scaffolding 'POLES' is introduced.

The problem of the optimal level of difficulty

Most teachers are familiar with the problem of deciding how difficult to make work that is set for learners. Some students always seem to want to be 'told the answers' (and some seem to think it is the teacher's job to simply provide information), but that is no good reason to acquiesce. Teachers set classes many varied tasks, not just to keep students busy, or to assess understanding (although that is useful), but because we tacitly know that learning is more likely to be achieved when learners are actively engaged.

Copying information from the board is often an 'activity' that implies that students are not actively learning, but this is not always the case. Good teachers can involve students in skilful expositions of a difficult concept area, with a certain amount of incidental copying (note-taking) along the way. However, it is generally recognised that most students do not actively process information when the task set is primarily of a copying nature. There are simple techniques to convert note-taking to a more active process (*eg* DARTs, which are discussed below).

Science, of course, is not just about learning facts (whether through copying or more active means), but involves learning how (and when) to apply the definitions, principles, models and theories that make up so much of what science is about. Useful learning involves developing the networks of meaning discussed in Chapter 3.

Worked examples may be given to get learners started, but students will only master the application of scientific ideas if they explore their use through the exercises set by the teacher. (At a higher level of understanding learners will be able to develop a mental 'toolkit' of concepts which they will use in solving problems where they have to select the appropriate tools and work their way to a solution.[1] Exercises, by contrast to problems, are used to practice the use of tools being acquired, and it will normally be clear to the learner which ideas he or she is expected to apply.)

Teachers become skilled at writing exercises, and – indeed – sometimes it may be too easy for the teacher to produce a set of practice questions for students. Some learners find security in being able to answer large numbers of very similar exercises. Often these are students who do not reflect deeply on learning (or indeed on the point of being in the classroom at all!), and who value a column of ✓s, and a mark of '20/20', more than having mastered new ideas. Brighter learners may well lose interest after a few similar exercises, and see little point in spending time in what they recognise as a largely algorithmic and repetitive exercise. They need something more challenging.

However, I am sure most teachers will recognise that it also possible to underestimate the difficulty of tasks set for learners (see the comments about the **Revising acids** activity discussed in Chapter 3). Something that is set as a quick exercise to reinforce a new idea, can become a major challenge that takes most of the class much longer than intended. In a subject such as chemistry such a problem can readily arise. For one thing the teacher brings to mind a wide range of pertinent and familiar background knowledge that learners may not recognise as relevant (see Chapter 4). The sheer subtlety and complexity of the subject – the wide range of models used for example (see Chapter 6) - can also overload learners.

RS•C

Clearly part of the problem facing any teacher is that of differentiation as within any class the learners will have a wide range of skills and abilities, levels of confidence and knowledge, and preferred learning styles. No lesson, or lesson material, will be pitched at an optimum level for each learner. Even ignoring this, and thinking of a single learner, setting activities with the right degree of challenge is a significant undertaking.

THE FAR SIDE®   By GARY LARSON

"Mr. Osborne, may I be excused? My brain is full."

A cartoon that always appealed to me showed a class at work at their desks, with one of the students with his arm aloft (Figure 5.1). The caption had the student asking to be excused, as his brain was now full. The humour that was intended derives from our familiarity with the fact that human brains seem to have an effectively infinite capacity to acquire more information. Our memories are not always accurate, and we do not always remember everything we might wish to, but they do not fill up!

Yet, I suspect, the reason I found the cartoon funny was that – like most good jokes – I knew it contained more than a 'grain of truth'. I empathised with the cartoon student. Whilst our memories do not reach their potential capacity, we all know that sometimes we are overwhelmed by information. Our ability to think clearly about a topic can become overloaded. This can either be because the information is too complex, or because it is simply arriving too rapidly. Most of us are familiar with being mentally 'lost' and needing to step back, 'clear our minds', and start again. Some learners probably find much of their experience of studying science to be of that nature!

RS•C

However, it is also common experience, that the same information may overload one person and not another, even when we judge them to be of similar ability. (And when the information content is different, the roles may be reversed.)[2] There are, then, two aspects of this phenomena that need to be explained; the experience of being 'overloaded', and its apparent variation between both individuals and context.

A limited register

Research shows that when people are asked to process nonsense information, they can only handle a very limited amount at a time.[3] Indeed, it has been suggested that our working memories can only handle 7±2 discrete items at once.[4] Most people could not hold a random 10 string of symbols (*eg* **f67H32md0w**) in mind unless they are able to find a way to chunk the information. Of course it is much easier to remember a meaningful string. (Even I could probably hold on to **0123456789** or **keithtaber**.)

The 7±2 rule (*ie* from 5 to 9 depending upon the individual) appears to be general, and to be a limitation of our cognitive apparatus, *ie* of the fixed structure of human brains. This provides a significant 'bottleneck' to our ability to process information: this working space can be considered to provide the channel between the enormous amount of sensory information available to each of us at any time, and the practically infinite store of memories.

In many cases this feature of the human brain will provide the 'rate determining step' in learning activities. The learner can only cope with material that can be processed in terms of a very small number of units at once. However, we are able to 'chunk' material so that what has seemed to be several different items may be counted as less. The extent to which we can do this is largely determined by experience. The email address **keith.taber@physics.org** consists of a string of 23 symbols (more than 7±2), but a reader with the right prior experiences (who was familiar with the general structure of email addresses, and the common domain endings {such as .com and .org}, and who already knew the name of the author), would be able to process the information as though it was comprised of many fewer than 23 units.

Thus the same information that will overload an individual who is not familiar with a topic, may seem perfectly sensible to another who is. This is true whether the topic is a football team, a boy band, Star Trek or aromatic chemistry.

If two learners of similar ability (and similar motivation to learn), are presented with the same information, but differ in background knowledge, they will cope differently. The same learner will make more of an explanation about a familiar topic than of another explanation of similar inherent complexity and logical structure about a less familiar topic.

Teachers, by definition, have much greater familiarity with the subject matter they teach than their students, and consequently it is very easy for teachers to underestimate the complexity of a learning task set for students.

Seeing chemistry at a different resolution

Elsewhere (see Chapter 4), I have discussed the problem of learners not having available in their minds the prior learning the teacher expects. This is an important problem, that may be approached to some extent by a conceptual analysis of the topic being taught (see Chapter 3). The teacher can specify the concepts upon which the new learning depends, and ensure learners have these ideas available before proceeding.

This is important, but may not be sufficient to ensure learners follow the teaching. It is not enough to check the right background is available, the teacher also has to ensure that the exposition provided, and tasks set, do not overload the learners' working space. This means that the teacher has to learn to perceive the conceptual structure at the resolution available to the learners. As most research into the learning of science has not (yet) focused on these issues, there is little specific advice to help teachers with particular topics. But with sensitivity and practice teachers can learn to see the complexity that

RS•C

the subject matter had, before years of thinking about chemistry resulted in much of it being neatly integrated into manageable chunks. The aim is to help learners move towards a similar level of conceptual integration, but this requires a major (mental) building programme.

Building needs foundations and scaffolding

Although the notion of 'constructing' knowledge may seem to be just a metaphor (see Chapter 10), the analogy between constructing a building and building an understanding of a subject, such as chemistry, is a useful one.

Just as a building needs firm foundations, so does learning. When the necessary pre-requisite knowledge is missing the structure can not be built (a null learning block), or at least does not match the architect's plans (a substantial learning block).

Perhaps for some buildings, solid foundations are sufficient. I imagine building a pyramid might be like this: each layer can act as the foundation, and access route, for the next. Most buildings are more tricky; although the final configuration should be stable, one has to pass through some unstable intermediate states before that arrangement can be attained. A partly erected building often lacks the structural integrity to hold together unless it has external support. Without such support the building programme becomes non-viable; floors can not be put in until there are walls to cantilever them, and walls cannot be reached because there is no floor to support the builder. In practice scaffolding is erected as a temporary source of support, until the building can progress to the stage where the scaffolding is no longer needed.

Knowledge construction can be seen to be closely analogous. Although one might imagine that a subject should be logically structured so that it can be built up brick by brick, many subjects can not be learnt in such a straightforward way (see the discussion of the nature of chemical concepts in Chapter 2). Perhaps mathematics might aspire to be a 'pyramid' subject, with each theorem absolutely standing on others, down to the foundations of initial axioms. Science, however, is not quite like this. Scientific concepts evolve, and become better elaborated. Chemistry is a subject built upon models (see Chapter 6) - which are often mutually supporting. Chemists invent conceptual entities to help make sense of their data, and each new concept (acid, element, oxidation, orbital, hybridisation) opens up new investigations which allow a finer grade understanding of the behaviour of chemical substances – and allow us to refine and redefine the theoretical entities themselves.

For the learner, chemistry has much of the same nature; the more that is understood about one set of ideas (eg oxidation in terms of electron transfer), the better one might appreciate another concept area (perhaps, acids as electron acceptors). As discussed in Chapter 3, chemical concepts need to be seen as part of a network of inter-related ideas. Both the development of chemistry, and the development of student understanding, may be seen as iterative processes, proceeding through spirals of increasing sophistication.

However, even if this was not the case, and chemistry could be reordered so that it could be taught as an entirely logical sequence of ideas, there would still be the problem of the learner's working space being limited to 7±2 items. Even though the learner may know which group of the period table chlorine is in, and what is meant by electronegativity, and what bond enthalpy is ... a given exercise calling upon this knowledge may seem to require too many different pieces of information to be mentally juggled at once.

It may be pertinent here to note the change in question styles over the years as the perceived purpose of the public examinations shifted from being a way of selecting a few, to a benchmark to be achieved by as many as possible. Although it is argued that there has been no significant drop in the level and amount of chemical knowledge required, only a few questions now require candidates to select and organise information into lengthy answers.[5] It is recognised that such questions are not very good at teasing out what most candidates actually know!

RS•C

Scaffolding learning (1): trust me I'm a teacher

It seems then that teaching a complex subject such as chemistry requires the teacher to do more that just present the material clearly and in a logical order. The teacher must also help the learner by supporting them when the working space is insufficient to hold all the relevant factors in mind at once. In explaining a new idea to the class the teacher will refer back to the relevant prior knowledge, but not just to show how it supports the present topic. The teacher's role is almost that of a confidence trickster, persuading the learner that certain points have been accounted for (as is indicated by the appropriate jottings on the board) and can be ignored – or just taken as given – for the moment.

The teacher is taking responsibility for certain parts of the logical support of a new idea, and asking the learner to focus on others. (This is a bit like a parent telling a child it is safe to try and swim, because the child is being supported and cannot sink. In both cases, some children require more convincing than others.)

Scaffolding learning (2): being in the zone

Ideas about teachers 'scaffolding' learning derive from the work of a Russian polymath called Lev Vygotsky.[6] Vygotsky wrote about learning (among many other things), and introduced the notion of the zone of proximal development. However, as (a) he did some great work, and (b) he had the misfortune to die young, and (c) it probably does not sound quite as clumsy in the original Russian, we should perhaps forgive his terminology. Even those who write about Vygotsky's ideas tend not to use the full term – it is quaintly known as the ZPD.

Vygotsky was very interested in the social side of the learning process (writing at a time when the Soviet system was still seen as a revolutionary idea), and realised that a learner is often able to achieve a great deal more when supported by an adult or more expert peer. This may sound obvious, but Vygotsky did not mean that the adult sometimes actually did the work for the learner.

Vygotsky had realised that learning is not an all-or-nothing process. He decided that intelligence tests that showed what a student could currently do unaided were not that useful to teachers. What was more informative was to see what the learner could not yet do alone, but could achieve with a limited amount of support, as this indicated where the child had the potential or readiness to develop new capabilities. The ZPD was the 'learning space' near enough to the child's current achievements for development to take place if suitable support was provided.[7]

This idea brings us back to the common teachers' dilemma of how hard to make a task. Make a task too easy and it is boring and does not bring about learning. A task that is too difficult will not be achieved, is de-motivating, and does not bring about learning either. The teacher needs to get the students working 'in the zone', and to provide the support that enables them to develop.

Scaffolding learning (3): what does scaffolding mean in practice?

'Scaffolding involves changing support over the course of a teaching session.'[8]

In practice, teachers have to be able to set tasks that students are not yet able to succeed in when totally unsupported. The teacher then provides the support, which is gradually reduced as the learner is able to master the ideas, until no support is needed. At this point the learner is able to mentally chunk material so that tasks that were too involved and overwhelmed their mental 'working space', are now perceived as having fewer separate components. Also, at this point, these tasks are no longer within the ZPD, as they are now within the child's capabilities. The child's ZPD has also moved on, as the recently acquired capabilities provide the basis for working towards new targets that are now just out of reach.

A large part of teaching involves oral exchanges, and it is often through these that teachers gauge the learner's readiness to tackle new challenges, and detect when to move in with support. (The answers to teachers' questions are often 'signposted' as they are designed to teach, more than question: after

RS•C

all the teacher generally already knows the answers.) These teaching skills develop naturally with experience, although explicit reflection on ideas such as scaffolding and 'the zone' may be helpful.

However, it is in setting written or practical tasks that the notion of scaffolding may be more obviously applied. Whereas classroom dialogue can be endlessly tweaked in real-time, worksheets and the like are presented to the class 'as seen', and calling for too many amendments in situ tends to undermine teacher authority and student confidence.

Planning to scaffold learning through written materials means thinking carefully about the conceptual and information processing demands of each task, and designing materials where the onus on the learner is gradually increased. One piece of good news is that some of the same ideas will be helpful when thinking about notching up the demands on individual learners as when planning to differentiate learning tasks and outcomes within a group.

Preparing teaching materials: DARTs

One of the points made above is that it is generally acknowledged that simply copying information does not lead to meaningful learning, as effective learning requires active processing of information. Of course, this is not the same as saying that learning never accompanies copying. As intelligent people who think about their own learning and thinking processes, teachers are among those who could probably think about the meaning of material whilst copying it. However, we all know that many students will either focus on the mechanical task of copying, or will allocate their minds (and perhaps their tongues) to some other activity whilst 'mindlessly' copying.

One alternative to asking learners to copy notes is to set them the task of making their own notes. However, to do this effectively is a skilled task which requires considerable practice, and teachers are commonly worried about the final results being incorrect or incomplete. As students' notes often form the basis for reference and revision, it is usually judged important that they are correct.

The purpose of DARTs is to provide a middle way which;

(a) gives a good chance of the learner having a full set of appropriate notes; and

(b) requires learners to think about the materials.

DARTs are Directed Activities Related to Text.[9,10] The simplest type of activities are passages with missing words, where the learner has to read the material to work out what the missing words are. (Sometimes Cloze procedure is used – removing every tenth words say – but it may be more effective to remove a number of selected key words. Variants include leaving the initial letter of the missing word, or providing the key words in a separate list.) Other activities can include labelling diagrams using information given in text or completing text using information given in diagrammatic formats.

It is possible to vary the degree of difficulty of DARTs to match students' needs. For example in a passage with words omitted, slow writers could be asked to fill in spaces on a sheet, when others in the group have to copy and complete the passage; and more words could be removed from a passage in the version given to the more able students in a group. There are a variety of DARTs type activities, but what they have in common is that they direct learners' attention to aspects of the text, rather than just copying. A number of the classroom resources provided in the companion volume include DARTs. The teaching exercises on **Precipitation; Elements, compounds and mixtures**; and on **Constructing chemical explanations** all have deliberate omissions which require students to think about the text they are reading. The less demanding versions of concept mapping activities discussed in Chapter 3 could also be considered as examples of DARTs.

The principle behind DARTs is not new. One variation is to provide a text passage, and have learners answer key questions about the passage in full sentences. The answers make up their notes, and this used to be called a 'comprehension' exercise!

Preparing teaching materials: scaffolding PLANKs and POLES

DARTs are designed to ensure that learners' minds are active when working on text. However, this alone does not ensure that the activity is effective at bringing about learning. As with all tasks teachers set, DARTs may still be pitched inappropriately. For example, a complete-the-missing-words-in-the-passage type activity might keep a group busy for twenty minutes without stretching most of the students.

If the task is too difficult the teacher will spot this when checking work (eg the omitted words are put back in the wrong places in the passage), but – as DARTs are normally designed so that the students' work is likely to be correct – it may be less easy to detect when tasks are undemanding.

In order to provide materials that help students develop their understanding, some of the ideas about scaffolding need to be taken on board. Students need to be provided with something more than just DARTs, they need to given tasks which enable them to develop their knowledge and understanding – DARTs which act as scaffolding tools (see Figure 5.2).

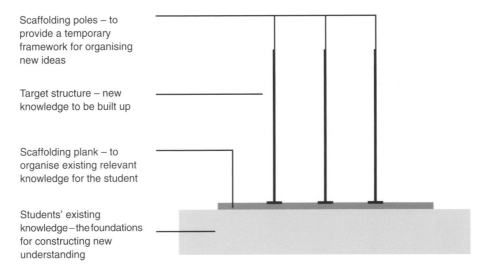

Scaffolding poles – to provide a temporary framework for organising new ideas

Target structure – new knowledge to be built up

Scaffolding plank – to organise existing relevant knowledge for the student

Students' existing knowledge – the foundations for constructing new understanding

Figure 5.2 Scaffolding student learning

There are two types of support that will help students to develop their understanding, and construct new knowledge. Firstly, even when students have available the necessary prerequisite knowledge for new learning they may not always be aware of which ideas are relevant (see Chapter 4). In addition, the limited register for processing information (see earlier in this chapter) makes it difficult for students to juggle the information so that they can use it effectively as the basis for developing new learning.

Secondly, the logical structure needed to develop the new ideas may exceed the processing capabilities of the student. Although each step in an explanation may itself be manageable, the overall structure may 'swamp' the student and seem much too complicated.

It follows that teachers can help in two ways. First, they can identify the necessary prerequisite knowledge, and not only be sure that students have covered the material, but that these ideas are marked out as relevant at the start of the new teaching episode. It may also be possible to organise the ideas for the students, into a form which will best facilitate the new learning. Secondly, the teacher can provide some form of partially constructed outline for the new knowledge, and make this available to the students as a guide for the new learning.

These two types of support may not always be clearly distinguished in practice, but it is useful to think of them as distinct types of support – taking the roles of providing 'horizontal' and 'vertical' support in Figure 5.2.

RS•C

Scaffolding POLES and PLANKs are designed to be components of the scaffolding which helps learners achieve at levels they can not reach unsupported. They enable them to practice, become familiar with, and feel successful with new ideas that they can then make their own. They are structures which offer support, but which will soon be outgrown.

PLANKs are PLAtforms for New Knowledge. Scaffolding PLANKs are presentations of ideas that are already available to students, but arranged in a form which aids the student in reorganising their knowledge to build up new ideas.

POLES are Provided Outlines LEnding Support.[11] Scaffolding POLES are provided by the teacher, and give a framework (outline) for exploring and succeeding in a concept area, that allows the learner to come to know about the topic. They lend support, because they are only to be relied upon whilst the learner is developing understanding and confidence in a topic.

Some DARTs may well be very effective PLANKs or POLES, but designing materials to help scaffold learning means taking additional requirements into account.

Principles behind POLES and PLANKs as learning materials

For learning materials to be considered as providing scaffolding, they should (individually, or as a set) meet the following criteria:

1. They must ask the learner to undertake an activity/task which is beyond their present ability if unsupported;

2. They must provide a framework of support within which the learner can be successful by relying on the structured support;

3. They must provide reduced support as the learner becomes familiar with the area, and is able to cope with increased demands;[12] and

4. They must result in the learner being able to undertake (unsupported) the activity/task which was previously beyond them.

The need for scaffolding PLANKs in learning chemistry

Earlier in this chapter the idea of 'seeing chemistry at a different resolution' was introduced. This simply means that the teacher has to try and see the complexity of a subject as it appears to the student. Even able and eager students are unlikely to have organised their chemical knowledge as effectively as the teacher – refining the organisation of knowledge can be a very slow process. This means that even when the teacher is convinced that the learner already knows of all the prerequisite knowledge needed for developing a new idea, the learner may not be able to readily access and order the information in the ways needed to build upon it. The teacher may need to help the students organise their knowledge.

Chapter 9 discusses an example of the type of learning difficulty that students may demonstrate. Research interviews found that some students may believe that when an ionic precipitate forms, there is an electron transfer to form the ionic bond. This finding has been reproduced in responses to a classroom probe included in the companion volume, **A reaction to form silver chloride**. So, for example, if solutions of silver nitrate and sodium chloride are mixed, then the precipitate of silver chloride may be considered to form, with an ionic bond between the silver and chloride ions. This may be explained (by students) as due to an electron transfer process:

'The outer electron in the silver transfers from the outer shell of the silver to the outer shell of the chlorine. This is called ionic bonding.'

Some students who made such responses had already, a few lines before in the same probe, demonstrated that they were aware that silver ions and chloride ions were already present in the reaction mixture. Yet in the face of an existing alternative conception (that ionic bonding always

RS•C

results from electron transfer to form ions – see Chapter 8), they did not effectively organise their knowledge about the species present in order to produce an explanation that was consistent with their earlier answers. Presumably the perceived complexity of the information available in this context prevents the student being aware of the contradiction in their answers.

An example of a PLANK

Consider, for example, that the idea of hydrogen bonding was to be introduced. There are a number of prerequisites that would be needed for the student to make the intended sense of the new concept.

A conceptual analysis (perhaps in the form of a concept map, as described in Chapter 3) might look something like that shown in Figure 5.3.

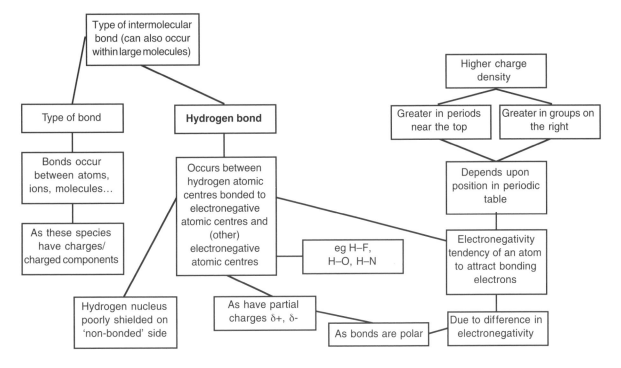

Figure 5.3 A suggested conceptual analysis for introducing hydrogen bonding

This particular scheme does not include all the ideas about hydrogen bonding that students may be required to learn (effects on boiling temperature, role in protein structure...), but shows the information being presented to introduce the new concept (shown in italics) and how this relates to the prerequisite knowledge that should be available to the students (but may often have been learnt in a more fragmentary way, and so may not be so well structured).

Once the teacher has made this analysis it may be used to plan the teaching. Figure 5.3 could be used to design a set of questions, or simply to provide set of teaching points that will be reiterated at the start of the lesson when hydrogen bonding is to be introduced. Figure 5.3 could also form the basis of a more specific PLANK for the students (see Figure 5.4).

RS•C

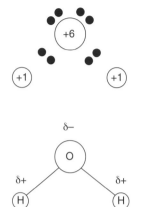

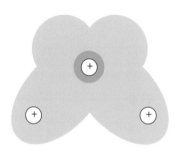

The water molecule

The diagrams are different ways of drawing the water molecule. They may help you to answer the following questions:

What do we mean by 'bond' in chemistry?

What exactly is the bond between an oxygen atomic centre and a hydrogen atom atomic centre in a water molecule?

How would you describe the average postition of the bonding electrons in the O–H bond?

Why are the bonding electrons not found half way (on average) between the oxygen atomic centre and the hydrogen atomic centre?

What type of bond holds the water molecules together?

Why are the bonds in a water molecule not strictly 'covalent'?

How would you describe the pattern of electron density (the shape of the 'electron clouds') in a molecule of water?

How well are the three atomic nuclei in a water molecule 'shielded' by the electrons?

Figure 5.4 A PLANK for organising prior learning

Figure 5.4 shows a student worksheet that might be used as an 'advanced organiser', to get students thinking about relevant ideas (bonds as attractions, electronegativity, bond polarity *etc*), and to organise these ideas into a suitable logical framework for learning about a new idea – hydrogen bonding.

Although some students may well find this activity sufficient to construct the idea that there will be forces (and therefore bonding) between different water molecules, the activity in Figure 5.4 does not explicitly lead students to construct this new knowledge. In order to help most students move beyond their prior learning, they will need an explicit input from the teacher. This could simply be a verbal exposition, building upon and developing the prior learning that has been highlighted. Alternatively, a specific learning activity could be provided – such as that shown in Figure 5.5.

RS•C

Interactions between water molecules

Activity: cut out the cards with the diagrams of water molecules.
Take three copies of the first type of diagram. Place them on a piece of white paper.
Imagine the molecules are in liquid water and are moving around near each other in the liquid.
Repeat this exercise with the different types of diagram.
Can you work out how the molecules will influence each other?
How will the molecules tend to become arranged?
Why does it take energy to pull the molecules apart?
If the liquid were to freeze, how might the molecules be arranged in the solid (ice)?
It has been suggested that there are bonds between water molecules in (a) ice, and (b) liquid water.
Explain whether you think this is correct or not.

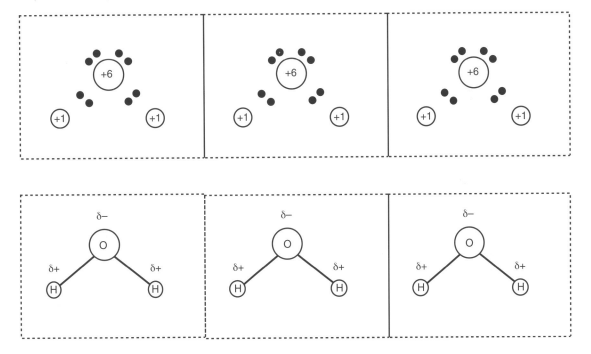

Figure 5.5 A scaffolding activity for developing new learning

The activity shown in Figure 5.5 provides a structured set of questions and an associated activity, designed to help learners construct a new understanding from a re-arrangement of their existing knowledge.

Whereas Figure 5.4 provides the 'advanced organiser' to 'prepare' the mind of the learner, Figure 5.5 provides a framework for building upon that preparation. The modelling activity provides a context to notice that molecules will attract – in certain configurations. The questions lend the outline of a logical argument to support the construction of new understanding (see Figure 5.6).

RS•C

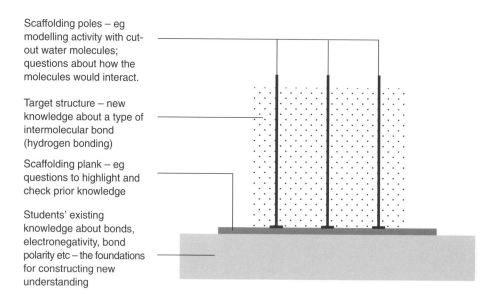

Scaffolding poles – eg modelling activity with cut-out water molecules; questions about how the molecules would interact.

Target structure – new knowledge about a type of intermolecular bond (hydrogen bonding)

Scaffolding plank – eg questions to highlight and check prior knowledge

Students' existing knowledge about bonds, electronegativity, bond polarity etc – the foundations for constructing new understanding

Figure 5.6 Scaffolding learning about hydrogen bonding

Examples of using scaffolding POLES in teaching chemistry

There are many places in a chemistry course where the teacher can provide a suitable outline to lend support to learners.

For example, some learners find mole calculations to be particularly difficult. A traditional approach would be to introduce the key ideas, provide a few worked examples, and then set some practice exercises.

Many teachers will carefully choose the examples, so that once a few questions have been successfully completed, some slightly more complicated examples are included. This approach will usually ensure the task is not too straightforward to challenge the most able (or most mathematically confident) in the group.

However, it is common experience that weaker learners often find even the most basic questions too difficult, and it is these learners especially that may need POLES.

The first level, beyond the totally worked examples, may be an example which is completely worked, apart from a one or two places where the student's input is required. These 'gaps' may be simply the result of numerical stages in the calculation (eg $12.0/3.0 = $ ____). This will help learners to see that the actual mathematical stages are quite straightforward, and rely only on the familiar arithmetic they use all the time. Once assured, they can start to focus on the chemistry.

In subsequent questions the level of support is gradually reduced. Each exercise should build on the previous either by being slightly more complicated (involving the relative molecular mass of a ternary compound rather than a binary compound; an additional significant figure in the data given), or by requiring an additional step to be undertaken by the learner.

This type of approach may be useful with both elementary mole calculations, and when introducing advanced students to the calculations involved in quantitative titrimetric analysis.

Similar approaches may be used with other types of calculations involved in chemistry, such as from Born-Haber cycles, electrode potentials, enthalpies of combustion, or oxidation numbers. In each such area, a 'script' can be produced, from which components can gradually be removed until the learner is working with nothing but the question data and a blank sheet.

RS•C

Scaffolding POLES in chemical explanations

Another area where learning may well need careful scaffolding is that in providing explanations. A common feature of chemistry lessons, and of chemistry examinations, is of using various models and chemical principles to provide explanations.[13] The range of explanations students are required to provide in an advanced course is quite large:

■ patterns in ionisation energies;

■ variations in lattice enthalpies;

■ shapes of molecules;

■ differences in covalent and ionic radii; and

■ differences in melting/boiling temperatures, *etc*.

In principle, a keen student should be able to learn the various principles involved, and so readily produce the appropriate explanations when needed. After all, once the principles are understood, it should be fairly obvious which ideas are needed.

Yet to many students these types of questions are quite mysterious, and the process of producing an explanation seems to be little more than guesswork – 'there was a question a bit like this last year, and the answer was hydrogen bonding, so I'll go for that...'.

Sometimes it may be quite frustrating to the teacher when the 'right' answer seems to be obvious to anyone who thinks about the question, but the learners shrug their shoulders and settle for their favourite catch-all, be it steric hindrance, entropy, d-level splitting or the presence of a lone pair. Keen, hard-working, students 'know' that questions about ionisation energy tend to need one of the stock answers ('it's in the p-orbital not the s-orbital', 'the electron is in a shell nearer the nucleus', 'the effective nuclear charge is greater', or 'it's due to spin-pairing'), but often seem to make a random selection.

Of course the teacher not only has a much greater familiarity with the subject matter, and a much better appreciation of how the different ideas fit together, but often also has years of experience of working through similar examples with successive classes.

Often the teacher is convinced that the learner 'knows' all the information needed to produce the right answer, that the student has learnt their notes, and only has to analyse the question logically. To the teacher the question can readily be answered without exceeding the 7 ± 2 capacity of the working memory: but not for the student. The student is trying to remember all the factors relating to patterns in ionisation energy, and think about the electronic configurations of the species specified in the question, and work out which orbital the removed electron was in, *etc* all at the same time.

The teacher may be carrying out the same set of logical operations, but is able to manage the process so that only the relevant points are kept in mind at each stage. As with the example of mole calculations, the student needs to be provided with a structure which helps them limit the amount they need to deal with at once.

Although this may seem like reducing chemistry to a set of algorithms, it is important to realise that the algorithms are only intended to lend support. As the student practices examples successfully (and therefore gains confidence as well as expertise) the outlines given with questions should be phased-out.

Modular explanations

Students may be given quite minimal tasks as they set out on learning a new skill. Most explanations required of chemistry students may be broken down into a discrete number of steps. The teacher may map out such explanations in the form of a simple schematic, such as a flow chart. The schematic may be used as the basis for POLES to be provided for students.

Consider the question: explain why aluminium has a lower standard molar first ionisation enthalpy than magnesium.

This is the type of item that a post-16 level (*ie* 16–19 year old) student would be expected to be able to answer in an examination or test, by producing a few lines of logical, coherent, and literate prose. The level of detail to be provided in the explanation will depend upon the amount of credit available. However, it is possible to prepare a schematic (see Figure 5.7) for the type of points that could be made.

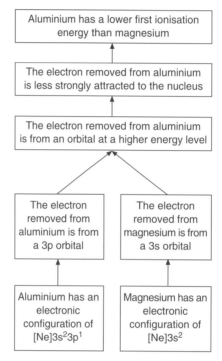

Figure 5.7 A schematic explanation

The value of the schematic is that;

■ it provides a logical analysis of the material;

■ it provides a starting point for producing scaffolding POLES; and

■ it demonstrates the constituent parts of the explanation.

The latter point is important because it will help students recognise the common elements that are often used as the components of explanations.

In using such schematics as the basis for scaffolding learning, there is a spectrum of possibilities that bridge between (at the highest level of support) providing the schematic, and then asking the learner to use it to prepare a short prose explanation of why aluminium has a lower standard molar first ionisation enthalpy than magnesium (*ie* a basic DART activity); and (when support is no longer needed) just setting the question itself. Two possible intermediate stages are shown below (Figures 5.8 and 5.9).

RS•C

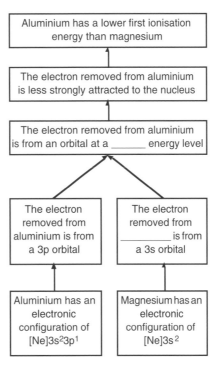

Figure 5.8 A DART of low level demand

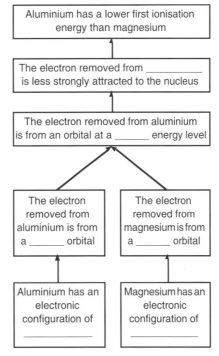

Figure 5.9 A DART of higher level demand

Figure 5.8 shows a version of the DART which provides nearly all the information, so that the missing words should be obvious from the logic and symmetry of the schematic. The second version (Figure 5.9) still provides a complete logical structure for the explanation, but requires the student to think much more about the direction of the logical relationships. Clearly the first version will be too simple for many students, but, similarly, some students would not immediately be able to cope with the demands of the second, and would need to 'build up to it'.

RS•C

The brightest students can be asked to develop the provided schematics further – for example incorporating additional factors such as the effects of increasing core charge and the amount of repulsion between electrons in the M shell, or extending the schematic to relate the electronic configuration to the elements' places (group and period) in the Periodic Table.

Providing a set of outlines (based on an explanation schematic) with different amounts of completion needed, for each of a range of exemplar questions about ionisation energy (or shape of molecules *etc*) is clearly a task that requires time and effort. However, once such materials are produced they should enable the teacher to match the task to the student to allow for both the range of abilities within a group, and to scaffold students at different stages to build up competency in providing appropriate explanations.

One of the classroom resources included in the companion volume, **Scaffolding explanations**, provides students with a set of questions, requiring explanations, similar to the example discussed above. Explanations set out as flow charts (with some missing elements) are provided to help support the student.

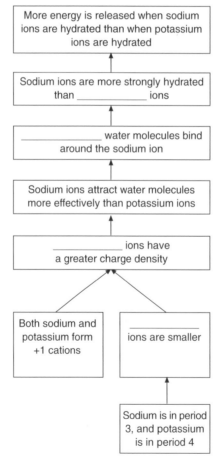

Figure 5.10 Explaining difference in hydration energy

For example, one of the statements that needs to be explained states that 'more energy is released when sodium ions are hydrated (390 kJ mol^{-1}) than when potassium ions are hydrated (305 kJ mol^{-1}).' One student, who was unable to give any explanation of this on the pre-test, **Explaining chemical phenomena (1)**, despite making attempts at other questions, was able to use the flow chart (see Figure 5.10) to construct an explanation:

'because more water molecules bind to sodium than to potassium because sodium attracts the water molecules more effectively as sodium ions are smaller and have larger charge density whereas both are +1 cations.'

RS•C

This, in itself, is hardly proof that this student has understood the ideas at a deep level, or will be able to reconstruct this explanation (or a related one) later when needed. However, being able to construct a valid explanation with the support of scaffolding POLES is seen as part of a process of gaining confidence and familiarity with the types of explanations used in the subject.

Notes and references for Chapter 5

1. K. S. Taber, An analogy for discussing progression in learning chemistry, *School Science Review*, 1995, **76** (276), 91–95.

2. For example, one study showed that although adults were better at recalling random digits than 10-year old chess experts, the youngsters outperformed the adults when asked to remember chess positions. See A. D. Pellegrini & D. F. Bjorklund, *Applied Child Study: a Developmental Approach (3rd Edition)*, Mahwah, NJ: Lawrence Erlbaum Associates, 1988, 128.

3. A. H. Johnstone, Some messages for teachers and examiners: an information processing model, in *Research in Assessment VII: Assessment of Chemistry in Schools*, London: Royal Society of Chemistry Education Division, 1989, 23–39.

4. G. A. Miller, The magical number seven, plus or minus two: some limits on our capacity for processing information, in *Psychology of Communication: Seven Essays*, Harmondsworth: Penguin, 1968, 21–50.

5. V. Barker, It's my party – and I'll cry if I want to!, *Education in Chemistry*, 2000, **37** (3), 72–74.

6. P. Scott, Teacher talk and meaning making in science classrooms: a review of studies from a Vygotskian perspective, *Studies in Science Education*, 1998, **32**, 45–80.

7. A. Drukovskis, The Zone of Proximal Development, at **http://chss2.montclair.edu/sotillos/_meth/00000014.htm** (accessed September 2005).

8. J. W. Santrock, *Life-Span Development (7th Edition)*, Boston: McGraw-Hill, 1999.

9. L. Bulman, *Teaching Language and Study Skills in Secondary Science*, London: Heinemann, 1985.

10. F. Davies & T. Green, *Reading for Learning in the Sciences*, Oliver & Boyd, 1984.

11. The acronym POLES was originally intended to stand for Provided Outlines Lending Epistemological Support – as epistemology is about how we come to know, *ie* our grounds for believing in something.

12. In other words, as the perceived demand of the original task seems less to the student.

13. J. K. Gilbert, K. S. Taber & M. Watts, *Quality, Level, and Acceptability of Explanation in Chemical Education*, 2001, available at **http://uk.groups.yahoo.com/group/learning-science-concepts** (accessed September 2005).

RS•C

RS•C

RS•C

6.   Chemical axioms

In this chapter some of the fundamental principles used in chemistry are considered. Research into learners' ideas about these key ideas are discussed, including the types of responses that were found when some of the classroom resources included in the companion volume were used in schools and colleges.

The chemist's creed

This chapter is called Chemical axioms because the principles discussed here may be considered to be the basic 'tenets' of chemistry: the key ideas upon which the whole subject is built. Learners who do not share these 'beliefs' – because they hold alternative conceptions – will find it very difficult to understand chemistry in the way teachers would want.

Material, energy and substance

One of the most basic discriminations in science is between energy and matter – although this is not a distinction that is always clear to students.[1] Although (since Einstein) it has been realised that the division is not absolute, it is still very important at the level of secondary and college science. Energy is known to be a difficult topic for learners, and younger students may not fully appreciate that heat (for example) is not a material substance.[2] Research also shows that children do not readily recognise that air and other gases are material – especially when there is no perceptible movement (such as a draught).[3] Something that cannot be seen, heard, felt, smelled or tasted (such as still air) may not seem to exist.

By the time students enter secondary school they should have overcome such problems, and recognise gases as material, and energy as something distinct from matter. Yet it is worth teachers of lower secondary groups being aware that a degree of confusion may still be present for some students.

A more common problem at this level is the way that chemists tend to use the term substance (or pure substance). This is a technical term in science. However, like many other words used in science, students will have heard the word 'substance' used with a more general everyday meaning akin to 'essence' or 'flavour'. To a chemist a substance has a definite chemical composition. To the learner any material may be considered a substance.[4]

To the chemistry teacher it is clear why sulfur, water and carbon dioxide are pure substances, but not wood, nor milk, nor air. Yet this distinction is based upon an appreciation of the composition of these materials at a sub-microscopic level that is not immediately available to the student. It is hard for the chemistry teacher not to think about any material in terms of its composition, but students are more likely to focus upon a material's appearance and its common uses.

Elements, compounds and mixtures

It is important to get learners thinking along 'chemical' lines, and lower secondary students are usually expected to develop an appreciation of the meaning of the terms 'element', 'compound' and 'mixture'. Yet – as will be discussed below – this distinction relies upon an appreciation of composition at two distinct and imperceptible levels.

Included in the companion volume is a set of classroom materials on **Elements, compounds and mixtures**. This includes diagnostic probes for (a) eliciting learners' understandings of these key terms and (b) applying their definitions to diagrams showing particles in examples of elements, compounds and mixtures. The examples are limited to molecular materials (*ie* not including giant structures) in order to avoid too much complication at this stage. The materials also include a teaching exercise.

The activity requires students to interpret diagrams representing molecules.

RS•C

A wide range of types of representation are used to show molecules in texts commonly used by students (see Chapter 7). Although this is potentially confusing for students it can also provide a context for teaching about the nature of scientific models (a topic considered in more detail later in this chapter). The various types of diagram used to show molecules each emphasise certain aspects of the molecules, and ignore others. Students need to appreciate that molecules do not 'look' exactly like any of these pictures, and that we select a suitable image to help explain or explore a particular aspect of the molecules.

For these particular classroom materials it was felt that modelling molecules as atomic cores in a cloud of electrons was suitable (see Figure 6.1). This type of representation would enable students to clearly identify discrete molecules in the figures, and to see how many types of atomic core were present in a molecule.

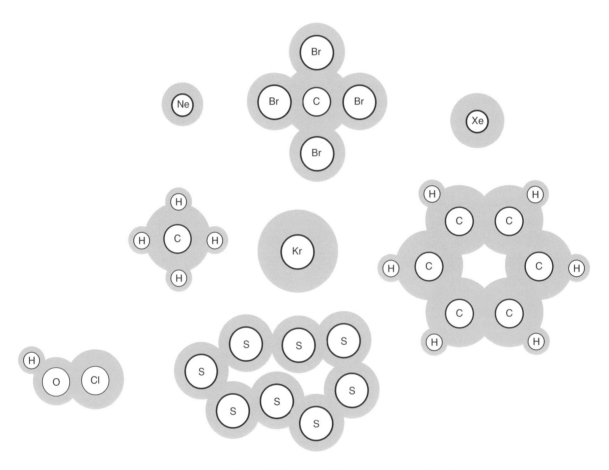

Figure 6.1 Representations of some molecules

When the materials were piloted in schools it was reported by teachers that students often had difficulty explaining the key terms adequately. (This is not be surprising in view of the difficulties teachers have in agreeing what terms like 'molecule' mean – see Chapter 2.) One teacher who was 'surprised at [the] confusion of key terms' suggested that teachers should pay more attention to the terminology used in the classroom. Another teacher who reported 'considerable uncertainty about definitions and the distinction between atom and molecule' went on to comment that this was still a problem with post-16 students.

A flavour of this confusion may be seen in the following definitions of a mixture from students in a class or 13–14 year olds;

'[A mixture is] something which is made up from two or more different kinds of atoms and molecules and compounds.'

RS•C

'[A mixture is] a substance with two or more different kinds of atoms [*sic*] which are not chemically joined together.'

'[A mixture is] a substance that contains two or more different atoms joined together.'

'[A mixture is] a mix of lots of different molecules in an atom.'

In some cases, students seem to be struggling to work out the scientific basis of the distinctions because they do not appreciate (a) that the distinction between 'atom' and 'molecule' does not parallel the distinction between 'element' and 'compound', and (b) they are not aware that they have to make discrimination at two levels (*ie* the number of types of molecules present in a sample, and the number of types of atomic core present in a single molecule).

In this same class of students;

■ tetrabromomethane (CBr_4) was classified as a mixture because 'there are two types of element' or 'different types of atoms in it';

■ sulfur (S_8) was considered a compound due to the number of atoms in the molecule;

■ a mixture of two hydrocarbons (CH_4 and C_6H_6) was classed a compound as there were 'two different types of atoms chemically joined together';

■ a mixture of two inert gases (Kr and Ne) was judged an element as 'the molecules are singular';

■ and the same mixture of two inert gases (Kr and Ne) was also judged a compound as there were 'two types of atom'; and

■ molecules of a compound of three different elements (HOCl) was classed as a mixture as there were 'more than 2 types of atom' or 'more than two types of element joined together'.

At least one student in this group seemed to have simplified the task of distinguishing the three classes in terms of the number of types of atom present:

Number of types of atom	Type of material
1	Element
2	Compound
3	Mixture

This student recognised the figure of S_8 molecules as an element ('only one type of atom') but considered both the mixture of noble gases and the mixture of hydrocarbons as compounds ('there is two types of atoms') and the diagram showing HOCl molecules as a mixture (due to the 'three types of atoms').

The probes and study task on **Elements, compounds and mixtures** deliberately include some examples of molecules that students would not normally have been familiar with, to ensure that they are applying their criteria for classifying materials, rather than just using recall. This includes using some examples of somewhat more complicated molecules, such as the benzene molecule (Figure 6.2).

RS•C

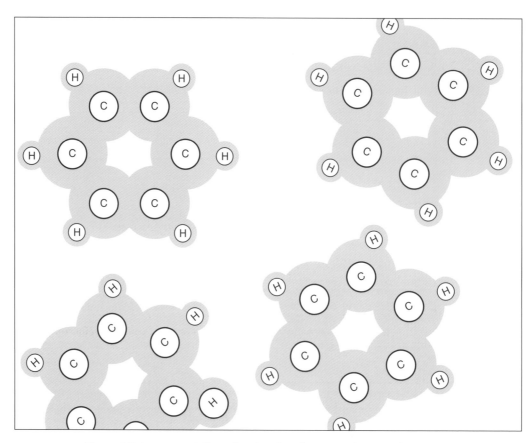

Figure 6.2 A representation of molecules of a pure substance (benzene)

One teacher reported that some of her (15–16 year old) students 'found the complex molecules (benzene) difficult to accept as a single substance'. In view of some of the responses reported above, it might be suggested that it is important that students do meet a wide range of examples of elements, compound and mixtures, and not just the simpler examples such as diatomic molecules and binary compounds. When students only meet a restricted range of examples, then they may well form alternative meanings for basic chemical terms as there will be insufficient counter-examples for them to have reason to question their own working definitions (see Chapter 2). As always, skilful teaching requires finding the optimum level of simplification for students to understand the key ideas without over-simplifying the science to be taught.[5]

The teaching exercise for **Elements, compounds and mixtures** is quite lengthy (with a good deal of reading) but includes spaces for students to demonstrate their understanding as they proceed. It was designed with the following principles in mind:

a) that it was important to break the task down into two discrete aspects – distinguishing a mixture from a pure substance, and then further dividing the single substances into those that were elements and those that were compounds;

b) that defining elements and compounds in terms of whether they could be 'broken down further by chemical means' was not very helpful, and a particle based model should be used.

The first point was very important in view of the comments made above about how learners may only have a vague idea about what is meant by the term 'substance'. These distinctions can only make sense to learners if they are asked to consider a particle model, and, even then, two levels of scrutiny are needed (see Table 6.1).

Looking at:	If the same:	If different:
The types of molecules	a single substance	a mixture
The types of cores in a single molecule	a molecule of an element	a molecule of a compound

Table 6.1 The summary table presented to students

Firstly the learner must consider the types of particles present (at the level of molecules) and decide whether one or more type of molecule is present. If the several types of molecule are present, then the material represented is a mixture regardless of whether the molecules present represent elements, compounds or both.[6]

Where only one type of molecule is present it is then necessary to look at those particles (the molecules) in more detail. Molecules comprise of one or more atomic cores enveloped by a cloud of electrons (see Figure 6.2). Students simply consider whether a molecule has a single type of atomic core or several types – and is therefore a molecule of an element or of a compound.

Defining elements and compounds

One of the classroom activities included in the companion volume, **Definitions in chemistry**, is designed to explore students' understanding of common definitions of basic chemical terms (*ie* element, compound, atom and molecule). This exercise is intended for students who have progressed beyond the stage where they will benefit from the classroom exercises on **Elements, compounds and mixtures** discussed above.

Some of the definitions included in this activity were derived from school textbooks or common reference books. Some of these definitions are dubious, ambiguous or just unhelpful (see Chapter 2). Also included in the exercise were some examples of definitions that were derived from students' comments, and these are lacking in accuracy or precision.

For example, the following definition of a compound was included: 'Is made of 2 elements mixed together'.[7] One student in a group of 14–15 year olds judged this as correct, though not helpful. This judgement was explained,

'I don't think it is very helpful for somebody learning science because it can be made up of more than one substance'.

A classmate judged this statement as correct and helpful,

'It is correct because compounds are 2 or more elements mixed [*sic*] together. An element is single, they make up compounds.'

Another classmate thought the definition was wrong and unhelpful, because,

'It could be more than two elements mixed [*sic*] together.'

These three responses highlight just how difficult these ideas are, and the language demands they place on students. The first comment seems to suggest the student misread the definition (although, as always with written responses, direct questioning would be needed to confirm this interpretation). The second ignores the limit on two elements in the definition, and agrees that a compound is a mixture. The final student has spotted that the wording of the definition does not allow ternary (or higher) compounds, but has also agreed with the notion of a compound being a mixture.

Models in teaching science

It is important that learners become familiar with using the molecular model of matter if they are going to be able to make sense of much of the secondary science curriculum.

RS•C

Research shows that students often have only a very limited idea of the way scientists use models.[8,9,10,11,12] At the start of secondary education many students will consider models to be little more than imperfect copies of an object, often made larger or smaller than the real thing. Scientists and science teachers actually use models in a number of ways that go well beyond this. For one thing scientific models focus on selected aspects of the system being modelled and deliberately ignore others that are considered irrelevant. As what is relevant changes with context, a particular model is designed for a particular purpose – for example to explain certain limited aspects of a phenomena. This means that it is quite acceptable to have multiple models which are mutually inconsistent, but which are used to explain different features of the same system.

Another aspect of scientific modelling is that models are mental tools, and so scientists do not always design models to be 'correct', but sometimes rather to test ideas. A model which 'does not work' when it is set up and explored – because it does not match the phenomenon being investigated – is like an incorrect hypothesis. Such a model can still be useful, as we can learn a lot from it by elimination. Students may not appreciate that these models are designed as exploration tools, and that scientists 'play' with the models in order to learn about the phenomena. In contrast, a student is more likely to assume that a model is intended to be as accurate a representation of 'reality' as possible, and to expect those models met in school science to have already been 'proved' correct.

Clearly it is useful for teachers to emphasise that models are hypotheses about the world which may be limited, partial, and even sometimes 'wrong'. This will help students to understand the limitations of models, to accept multiple models, and to appreciate a little more about the nature of science itself.

One of the most important models (or perhaps, collections of models) used in chemistry is the molecular model: the idea that matter is not continuous but made up from discrete particles that are much too small to be seen.

The particle model of matter

Although the molecular model is a central idea in science, and is very familiar to chemistry teachers, we must not be complacent about how problematic particle ideas are for students.[13] Research shows that students do not tend to spontaneously use relevant particle ideas in explaining chemical phenomena.[14] This is understandable when the difficulties that students face learning the model are considered.

The word 'particle' is not very helpful, as it is used for small (yet still visible and macroscopic) particles such as salt grains or dust specks, as well as a collective term for molecules, ions etc. Research shows that this leads to confusion for many students.

A more intractable problem is deciding what the particles are, from which all materials are made. are. Teachers like to have a single term that they can use when introducing the molecular model to students, but this can lead to rather imprecise use of language – as in the following classroom observation,

'the teacher conjured up an image of diffusion in solutions by referring to blue copper sulphate 'atoms', and colourless water 'atoms' wriggling slowly past each other at the junction of the two layers'[15]

For many substances it is quite correct to refer to the particles as molecules. This would be appropriate for sugar, water, sulfur, oxygen etc. However, not all materials are molecular. In diamond it is quite possible for the 'molecule' to be a visible entity. In metals such as iron, and salts, such as sodium chloride, there are no molecules.

This is not a pedantic point, as learners who are taught that 'everything is made of molecules' will expect iron and sodium chloride to contain molecules, which can be a problem at a later stage in science learning (see Chapters 7 and 8).

RS•C

One alternative is to use the word 'atom'. It is common for chemistry books to suggest that everything is made of atoms (see Figure 6.3), but this is also problematic. Very few substances (*ie* the noble gases) are actually composed of atoms per se. In molecular materials the discrete particles present are molecules and it is not helpful if students assume that something like oxygen comprises of oxygen atoms (see Chapter 10).

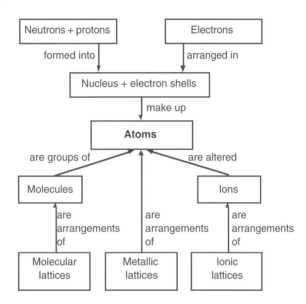

Figure 6.3 Seeing atoms as the building blocks of matter

Although we might commonly say 'a molecule of water contains two atoms of hydrogen and one of oxygen', it can be argued that this is not quite correct either. In a water molecule some of the electrons (the bonding electrons) are not associated with a single nucleus, and should not be considered part of a particular atom. This may again seem pedantic, but students have been found to take the 'molecule contains atoms' idea literally, and to assume that each electron 'belongs' to and is 'part of' only one of the atoms. This leads to false ideas about the interactions in the molecule (with electrons only being attracted to their own nuclei!) and about bond fission always being homolytic (because a bond always breaks so the electrons go back to 'their own' atoms!)[16]. Similar ideas also lead to common misconceptions about ionic bonding (see Chapter 8) and to problems understanding precipitation reactions (see Chapter 9).

The atom is also an inappropriate label for the particles in salts such as sodium chloride or metals such as iron, both of which are better understood to contain ions – with, in the case of metals, delocalised or lattice electrons which are often referred to as a 'sea' of electrons (see Chapter 8). Even in the case of diamond there are no discrete atoms present, although again we find students may sometimes assume that diamond does comprise of separate carbon atoms (see Chapter 7).

It would of course be correct to say that all matter consisted of protons, neutrons and electrons, and we would want students to learn this during their secondary careers, but this is not considered an appropriate treatment at an introductory level.

All the structures that students are likely to meet in chemistry can be described in terms of arrangements of atomic cores and outer shell electrons (see Figure 6.4), but this is a way of conceptualising matter that will take time to be acquired once the simpler, if inadequate, 'everything is made of atoms' notion is accepted.

RS•C

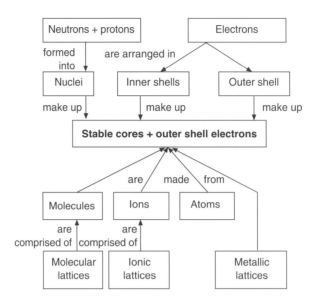

Figure 6.4 A more scientific, but more complex, model of the building blocks of matter

It is difficult to provide definitive advice in this topic as there does not seem to be an obvious solution to the problem. All of the alternative approaches are fallible, and are likely to lead to students acquiring inappropriate ideas unless presented very skilfully.

Probably the best approach is to refer to 'molecules' in introductory work, and use suitable examples, but to make it clear that the particles in some materials are slightly different. It is not necessary to give any details (although the example of metals might be mentioned), but it is sensible to emphasise to students that they are being taught an important, but incomplete model, which will be developed later in their education. At a later stage of the students' education, when the ideas of chemical structure and bonding are being introduced, it may actually be more sensible to first consider the simplest case of metals, and to teach about solid structures which are arrays of simple molecules last (see Chapter 10).

Defining atoms

One of the classroom activities included in the companion volume, **Definitions in chemistry**, is designed to explore students' understanding of common definitions of basic chemical terms (*ie* element, compound, atom and molecule) – something that is inherently problematic (see Chapter 2). As discussed earlier in this chapter, this resource asks students to comment on definitions taken from books and from students' comments.

For example, the following definition was given by a student setting out on a post-16 chemistry course,[17]

'an atom is the simplest structure in chemistry. It contains a nucleus with protons and neutrons, and electrons moving around shells.'

This was included as an item in the **Definitions in chemistry** exercise. One student in a class of 14–15 year olds was not sure if this was a helpful definition as 'the electrons move around in shells, not around shells'. A classmate judged it as a correct and helpful definition: 'It's to the point, clear and fairly easy to understand'. Neither of these students questioned the idea that the atom was the simplest structure in chemistry. Indeed, another classmate agreed with this point, whilst pointing out that the nucleus also had structure,

'atoms are simple structures, the nucleus is made up of protons and neutrons and electrons orbit it in shells.'

RS•C

This same type of internal contradiction was present in another of the definitions included in the exercise. This was also based on the definition provided by a student commencing a post-16 chemistry course. For this student, the atom was the

'smallest particle that can be found. Made up of protons, neutrons and electrons'

Yet, this contradiction is not obvious to some students. When this item was considered by students working through the **Definitions in chemistry** activity, one of the 14–15 year olds judged this as a correct and helpful definition, and explained,

'They are the smallest possible thing and they are made up of protons neutrons and electrons.'

As discussed above, there is a tendency for the term atom to be used in introductory science as a catch-all term for atoms, molecule and ions, and students readily accept the 'everything is made of atoms' mantra, even though this is (at best) a gross over-simplification. These examples show that some students are also happy to accept notions of atoms being the simplest structure possible and the smallest particles that can exist – even when they 'know' they contain simpler structures and smaller particles. That such contradictions do not seem to worry these students is another indication of how little sense many students make of our molecular models of the world.

Student confusion over molecules and ions

Students often find it difficult to think in terms of the images of molecules and the like that seem to come so readily to experienced teachers. For example, one of the diagnostic probes in the companion volume – part of the materials on **Precipitation** – asks students about the species present in sodium chloride solution. It become clear during the piloting of this resource that some 14–16 year old students found it very difficult to work out which molecules and ions were present in solutions.

The students were presented with a diagram of solid sodium chloride (see Figure 6.5) which was accompanied by text informing them that 'the particles in sodium chloride are sodium ions (Na^+) and chloride ions (Cl^-)'.

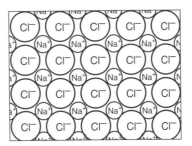

Figure 6.5 Particles in solid sodium chloride

The students were also provided with a diagram representing molecules in water (see Figure 6.6), which was accompanied by text informing them that 'the particles are water molecules'.

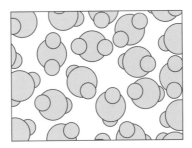

Figure 6.6 Particles in liquid water

RS•C

It is worth noting here that there are genuine problems in trying to represent liquids in simple particle diagrams. This representation could be criticised as exaggerating the space between molecules. However, as liquid molecules do not form 'layers' a more accurate diagram would either have to show complete molecules from different 'depths' in the liquid (which would overlap and complicate the diagram), or would need to be a true two dimensional cross section of a slice through the liquid – so that different molecules would be sectioned at different points and would all seem to have different shapes. As discussed above, all representations are models of selected aspects of what is represented – and opportunities should be taken to raise the limitations with students.

In the **Precipitation** probe, the first task set to students is to list the species present in sodium chloride solution. The students are asked:

'Sodium chloride dissolves in water to give sodium chloride solution. What particles (such as particular atoms, molecules, ions) do you think are present in sodium chloride solution?'

Although the probe was primarily concerned with students' ideas about what happens when an ionic precipitate formed, it is clearly necessary for them to appreciate which species are present in the two solutions which are mixed in the reaction. The first questions were designed to check students were aware which species were present, and to focus their thinking on this as a relevant aspect of the process (see the comments on scaffolding PLANKs in Chapter 5).

Clearly the main species present in sodium chloride solution are sodium and chloride ions and water molecules. In view of the degree of cueing (or scaffolding) provided by the diagrams and accompanying text, it is not surprising that many students were able to give the correct response.

However, by no means all students were capable of working this out. Among the responses suggested in one class of 14–15 year olds in one school were the following:

■ sodium ions; chloride ions; hydrogen ions; hydroxyl ions

■ Na^+ ions; Cl^- ions; water molecules; sodium chloride ions

■ sodium ions; chloride ions; hydrogen ions; oxygen ions

■ sodium atoms; chlorine atoms

■ molecules of NaCl and H_2O molecules

■ sodium ions; chloride ions; hydrogen molecules; oxygen atoms

■ sodium ions; chloride ions; hydroxyl ions.

Although some keen students could possibly be aware that some ionic association may occur, and that water has a degree of dissociation, this can not explain responses including NaCl molecules instead of ions, or hydrogen and hydroxyl ions but not water molecules. These students were generally struggling to report what would be present when solid sodium chloride (which they were told consisted of sodium ions and chloride ions) dissolved in water (which they were informed comprises of water molecules).

Other students in this class seemed to be struggling to give an answer in particle terms at all, suggesting that the solution contained hydroge [*sic*]; oxygen; sodium; chlorine or 'a solid'.

Defining molecules

One of the classroom activities included in the companion volume, **Definitions in chemistry**, is designed to explore students' understanding of common definitions of basic chemical terms (*ie* element, compound, atom and molecule) – something that is inherently problematic (see Chapter 2). As discussed earlier in this chapter this resource asked students to comment on definitions taken from books and from students' comments.

One of the definitions of 'molecule' included in the activity was derived from the comments of a student, who had defined a molecule as 'formed by two atoms bonding together'. That this only

RS•C

allows diatomic molecules might seem an obvious limitation. Again, however, students may not recognise this problem. One student judging this as correct and helpful explained that it is 'quite easy to understand [and] doesn't complicate matters later on'. Even this uncomplicated definition elicited some evidence of students confusing the molecular model with the macroscopic phenomena. One classmate thought that this definition was 'correct, but not very clear. [As] It doesn't explain the different between compounds and molecules'. A compound would be said to be formed from elements, not atoms, of course. Another student in the group thought that the definition was both correct and helpful because it was 'like water which is made of 2 elements' – although a molecule of water, of course, has three atomic centres.

Once again, there is evidence of students confusing the most basic ideas in chemistry, and not being able to clearly distinguish between the level of substances and their representation through molecular models.

Applying the molecular model

Research shows that even when learners may seem to have acquired the 'everything is made up of particles' principle, there can be serious flaws in their understanding of what this is meant to imply. For example learners may suggest that the substance is found between the particles – so in water there are water particles with water in-between them! Alternatively they may just assume there is air between the particles (after all, it is said that 'nature abhors a vacuum'!)[18] A related problem occurs when students think about the bonds between the particles in a material – which they often imagine as material links such as springs or elastic (see Chapter 8), showing that they have not really taken on board the full implications of the molecular model.[19] These types of ideas show just how abstract and far-removed from everyday experience the molecular model is.

Even when students have 'got' the particle idea, and realise that the particles are all that is there (not particles embedded in the substance, or separated by air), they may fail to appreciate the value of the model to science. The particle model is so successful because a large number of macroscopic properties of substances can be explained in terms of the conjectured properties of the systems of particles (see Figure 6.7).[20]

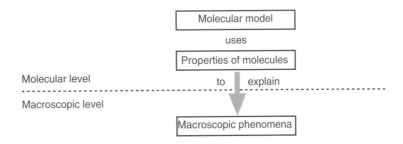

Figure 6.7 How science uses the molecular model

Without appreciating this point, the rationale of the molecular model is lost. And yet there is a great deal of research that shows that students commonly do fail to understand this. Although learners will learn to talk about particles when explaining macroscopic phenomena, they often simply transfer the property to be explained to the molecular level (as in Figure 6.8).

RS•C

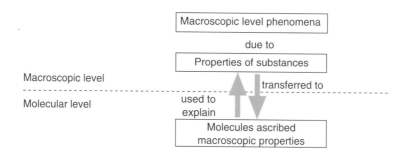

Figure 6.8 How many students apply ideas about molecules

There are many examples of this that have been reported. So materials are said to expand on heating because their particles expand, for example. In science we teach students that thermal expansion can be explained on a particle model (Figure 6.9).

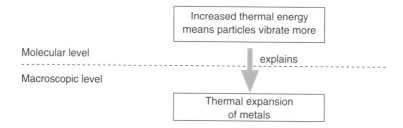

Figure 6.9 An example of how students are expected to apply molecular ideas

Students generally come to accept this model, even though it has an obvious logical flaw (if the particles are vibrating synchronously the increased vibrations need not change the overall volume!) It is only later (for those who study post-16 level physics) that a more comprehensive version of the model is provided (Figure 6.10):

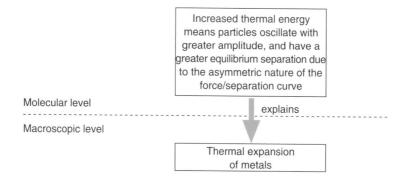

Figure 6.10 A post-16 level explanation of thermal expansion

Although students seldom query the simplified explanation given in Figure 6.9, this is more likely to be a result of their failure to visualise the model, rather than finding the argument convincing. The type of particle-based explanation that many students themselves suggest follows the pattern described above (see Figure 6.8), of transferring the property to be explained, in this case expansion on heating, to the particles in the model (see Figure 6.11).

RS•C

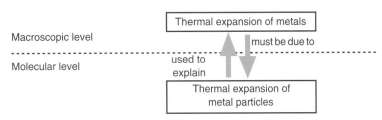

Figure 6.11 A common way for students to explain thermal expansion

This type of approach shows how students commonly miss the point of having a molecular model.

The idea that 'materials expand on heating because the particles expand' is regarded as a common misconception. This is a little unfortunate for students as if the metal (or other material) is made up of particles with 'nothing' between them, and if the bulk material gets bigger on heating, then on average each particle must clearly take up more space. This would seem to suggest that they get bigger! This could be considered to be at least as sensible a suggestion as the accepted response (Figure 6.9) at this level.

Even though I have sympathy with students in this particular case, there are many other examples where this approach is clearly unhelpful. Individual molecules may perhaps be considered to have a smell, and some are coloured (although colour is often a property of higher levels of organisation), but there are many properties of materials that are due to the arrangement of, and type of bonding between, particles rather than just the particles themselves.

Among the resources included in the companion volume is a classroom probe, **Iron – a metal**, which asks students to judge the truth of a number of statements about the properties and structure of iron. Some of the items test whether students will find acceptable explanations of metallic properties which transfer the macroscopic property to constituent particles (*ie* as in Figure 6.8). When this activity was tackled by one group of 14–15 year olds, it was found that a majority of the students agreed with a number of explanations of bulk iron properties in terms of the individual atoms rusting, conducting electricity and reflecting light:

■ The reason iron rusts is that iron atoms will rust if exposed to damp air.

■ Iron conducts electricity because iron atoms are electrical conductors.

■ An iron atom will reflect light, and so freshly polished iron shines.

Chemical and physical change

Another basic distinction which 11–14 year old students are expected to develop is that between chemical and physical changes. Again we find a fundamental distinction which is problematic for teachers and learners. In Chapter 2 we saw how this basic distinction becomes difficult for teachers when they are asked to classify changes such as dissolving, or driving water of crystallisation from hydrated salts.

Although the difference between these two types of change is often considered significant, it is not easy to provide learners with simple rules for making the distinction. Two common types of definitions are used (Table 6.2).

Physical change	Chemical change
easily reversed no new substance	not easily reversed new substance formed

Table 6.2 Key criteria for distinguishing chemical and physical changes

RS•C

Clearly chemical changes can be reversed, although sometimes only under extreme conditions. However, those chemical reactions with a very low Gibbs free energy change may be quite readily reversed. Conversely, mixing is considered a physical change, but separating some mixtures (such as in crude oil) may not appear to be 'easy' to students. From the teacher's perspective, of knowing which changes are considered ('meant to be') physical and which are considered ('meant to be') chemical, then the judgement of whether reversal is easily achieved may well sometimes be prejudiced!

Similarly, deciding whether a different substance is formed during a change will be easier for the teacher, calling upon privileged information that may not be available to the students.[21] It is a fundamental tenet of chemistry that – for example – steam, liquid water and ice are the same chemical substance despite having different physical properties. (In principle they have the same chemical properties, but this is not easily shown: does ice react with sodium without melting? When magnesium is reacted with water there is no obvious reaction until the water has boiled.)

A change of state is a physical change as the same substance is present before and after the change. And yet, at the start of their chemical education, students only have it on trust that ice and steam are both forms of water, yet (for example) rust is not the same substance as iron. Again these distinctions only become 'obvious' when viewed from the 'molecular' level. Steam, liquid water and ice all contain the same basic particles – water molecules. This means that it can be argued that it is often actually easier for students to judge whether a change is physical or chemical on the basis of particle model diagrams than by looking at the process itself!

Two other criteria that could be introduced here (Table 6.3) are equally problematic.

Physical change	Chemical change
easily reversed	not easily reversed
no new substance	new substance formed
no bonds formed / broken	bonds broken and / or formed
modest energy change	large energy change

Table 6.3 Criteria for distinguishing chemical and physical changes

The idea that in a chemical change bonds are broken and/or formed seems sound. Yet it is not so clear that a physical change does not also involve bond breaking. At lower secondary level we probably only think of covalent, ionic or perhaps metallic bonds as being chemical bonds. Yet, later, we may wish students to learn about hydrogen bonding, van der Waals forces, and solvent-solute interactions. Research shows that learners may find it difficult to expand their definition of chemical bonds to include these categories (see Chapter 8), and so it probably unwise to restrict our meaning of 'bond'.

Clearly physical changes such as changes of state, do involve the breaking and forming of bonds – this is why ice, water and steam are so different despite being made up of the same water molecules.

The energy change criteria is just as problematic: although some chemical reactions (*eg* the hydrogen/oxygen explosion) may obviously involve large energy changes, and some physical changes (*eg* evaporation of a volatile liquid) may seem to involve little energy transfer, it is possible to suggest counter examples. The rusting of iron seems no more an energetic process than evaporation, and students are warned about the dangers of scalding themselves on steam, because of the large amount of energy released when it condenses.

Some teachers feel that the chemical / physical change distinction is an unhelpful one which should not be taught. However, at the moment it is a required part of the science curriculum for many students, and so must be covered in class. The best that can be said is that these concepts have 'fuzzy' boundaries (see Chapter 2).

RS•C

Chemical and physical changes are not categories that obviously exist in nature. It is chemists who find it useful to impose these artificial categories upon the wide range of changes that can occur to, and between, substances!

One of the class resources provided in this publication, **Changes in chemistry**, is a probe to test whether students can recognise which changes are considered to be chemical and which are said to be physical. In view of the problems discussed above, this probe provides 'before' and 'after' diagrams based on particle models to help students make their decisions.

Although it is possible to avoid some of the difficulties in this way, and by limiting examples to clear-cut cases, it seems sensible to emphasise to students that:

■ the distinction between physical and chemical change is just one that chemists sometimes find useful;

■ although they will only be expected to judge examples with clear answers, some changes they will meet do not fit easily into one category.

When the probe was piloted in schools it was found that students were often able to demonstrate acceptable ideas. So freezing liquid nitrogen was a physical change because;

'no chemical bonds have been broken...'

'the particles are the same they have just been restructured'

'the chemical formula has not been changed...',

and

'...it is the same substance...'.

Burning magnesium was considered to be a chemical change,

'because the Mg and O has bonded they have a chemical bond and can not be changed back. Magnesium oxide has been produced a new substance.'

However, some of the problems with the chemical/physical distinction were reflected in students' responses. This may be seen in the responses from some students in groups of 13–14 year olds. For example, although most students recognised that combustion of magnesium was a chemical change, there were some dissenters. Some students were able to see this as a reversible process. One focused on the perceived increased activity of the particles,

'the molecules have been heated and so they are very active but if you cool them down they will be able to settle.'

Other students thought that

'magnesium oxide can be changed back to magnesium and oxygen'

and one suggested that

'you could do a displacement reaction to get the oxygen away from the magnesium oxide. You would be left with oxygen and magnesium.'

The case of sodium chloride dissolving produced a range of responses from students seeing this as a chemical process. It may be relevant here that students often use the same terminology for the process of a solute dissolving in a solvent, and, for example, a metal reacting with ('dissolving in') an acid.[22] A number of students argued along the lines that 'you are unable to turn the substance back to it's original form as it is mixed in with the water.'

Whereas a simple distinction based upon reversibility implies a clear demarcation, what we find when students try to apply the reversibility criterion is that some see chemical changes as reversible (suggesting chemical methods that are 'not allowed'), and others consider physical changes as

RS•C

irreversible (presumably not considering a technique such as distillation as 'allowed'). This should not be surprising when we realise that the reversibility criterion is inherently tautological, *ie*,

■ a chemical change cannot be reversed except through a chemical change;

■ a physical change can be reversed through a physical change.

When it is spelt out in this way it is clear that such a criterion needs to be learnt through familiarisation with conventional examples, and can not be based on definition (*cf* the comments on defining concepts in Chapter 2).

Some students gave other arguments for considering dissolving of an ionic substance as a chemical change. Some noted that bonds had been broken in the sodium chloride,

'The Cl & Na are no longer bonded together and are able to move around separately from one another.'

'the Na and Cl have been split up and the atoms are more widely spaced and in a mixture the bond between the Na + Cl has been broken'

and one student suggested that new bonds had been formed (which is technically correct, as solvation had occurred),

'the different atoms have bonded together & formed a new pattern.'

The **Changes** probe was also undertaken by a group of High School students in Greece. Although there could be some doubts over the additional demands of undertaking the probe in a foreign language (*ie* English), it is interesting that most of the group considered the dissolving of sodium chloride to be a chemical change (11/14, with the other 3/14 considering this change to be physical). As with the British respondents, there were some attempts to justify the classification in terms of pertinent criteria,

'some sodium chloride is added to a beaker of water and left to dissolve and so the molecule are connected'

'water (H_2O) divides NaCl into its components Na, Cl so it's substance change'

Although student responses did reveal some genuinely alternative ideas about the processes discussed (such as weight changing during a physical change, or a chemical change being one where 'you cannot actually see the change'), this particular probe might be considered as most useful for diagnosing students' inappropriate application of the fuzzy set of criteria used in the science curriculum for distinguishing chemical and physical change.

Conservation in chemistry

Some of the most basic principles in science are conservation laws. In physics classes students learn about conservation of energy, and this principle apples in chemistry as much as elsewhere. At post-16 level, students will be taught that mass can be considered as a form of energy, but at secondary levels energy and mass are considered as separate.

The conservation of mass is a very central and basic idea in chemistry. Its application is (again) tied closely to particle ideas. The fundamental particles from which substances are composed have fixed mass, and the mass of a sample of material is the sum of the masses of all the constituent particle masses.

Two key teaching points, then, are that during a chemical reaction (or during a physical change):

■ the same fundamental particles are present at the end as at the start; and

■ the total mass has not changed.

The particles are rearranged, but the new configuration is made up of the same fundamental particles, and therefore has the same total mass, as before.

RS•C

Of course, the 'particles' here are not molecules, but the atomic constituents. As the mass of the electrons is less than 0.1%, and as there are no nuclear transformations being considered, the mass in chemical reactions may be considered to 'follow' the atomic cores, or nuclei. (So we can use relative atomic masses to calculate reacting masses, and to find percentage yields etc.)

One of the classroom exercises included in the companion volume, **Mass and dissolving**, presents an example of a type of change where students should realise that mass must be conserved: dissolving.

It might be thought that conservation of mass would be more obvious in the case of a physical change than in a chemical reaction (if dissolving is considered a physical change – see above). However, research suggests that learners do not always understand what happens to the solute when it dissolves.[23]

The classroom exercise provided is designed both to elicit students' ideas about what happens to the mass present when a solution is formed, and also to challenge their thinking. The exercise requires students to predict the mass of a liquid and solute (of given masses) when first mixed, and when the solute has dissolved. They are also asked qualitative questions about the examples:

1. Sugar/water: the students are asked to explain where the sugar went when a solution forms.

2. Copper sulfate/water: the students are asked about the colour change (to focus their thinking on a property of the solute now transferred to the solution), and then where the copper sulfate went.

3. Sugar/water: students are now provided with particle diagrams of the solvent and solute, and the resulting mixture, and asked why the liquid tastes sweet (to focus their thinking on both the property of the sugar transferred to the solution, and to provoke them to think about the particle model).

4. Salt/water: students are asked where the salt went.

The design of the probe is intended to help teachers identify 'non-conservers' – pupils who do not realise that the solute is still present in the solution, and therefore its mass is also present – and then challenge these individuals to think about why the 'water' has new properties once the solute is no longer visible.

When this probe was piloted in schools it was found to be effective at diagnosing students who do not appreciate that mass is conserved on forming a solution. Most students know that when a solid is first added to a solvent the weight will increase accordingly. Many students will be aware that the weight of the solute will continue to be registered after the solution is formed:

'It dissolved into the water but was still there so the weight was still there.'

However, a significant proportion of students assume that when the solid is added to a solvent (and can still be seen at the bottom of the beaker) the mass of the solute will register, yet think that once the solid cannot be seen then the weight of the mixture will return to what it was before the solid was added.

This might be seen to be consistent with literature reports that younger students may simply consider the solute to have 'disappeared' once it can no longer be seen.[24] However, when the materials were piloted with 11–13 year old students for this project, it was found that many of the 'non-conservers' were aware that the solute was still present, albeit not visible:

'The sugar dissolved into the water which made it look like the sugar has disappeared.'

'The salt dissolved into the water and is still in there but cannot be seen'

'The copper sulfate dissolved into the water but it hasn't gone anywhere, it just can't be seen'

Clearly, for these students, knowing that the solute is still present does not necessarily imply that its mass will be registered. One explanation for this is simply that these students do not conserve mass, *ie* do not recognise a need for the solute to continue to have mass when not visible – after all if some

RS•C

of its attributes (surface, shape, grain) are no longer perceptible, then perhaps weight similarly 'dissolves' away.

However, responses to the exercise suggest that some of the non-conservers are aware that properties such as taste and colour may be retained ('the sugar has dissolved all in to the water but you can still taste it because there is [sic] millions of sugar particles in the water'; 'it has turned blue because the particles has [sic] spread and you can't see the lumps of it'), and it may be that some students are assuming that the dissolved solute has buoyancy and will not register any weight.[25] (Experience with buoyant objects in baths and swimming pools appearing weightless might suggest this.) Whatever the reason, the exercise does provide an opportunity for teachers to diagnose this belief and challenge it. The questions in this classroom exercise can be readily replicated on a top pan balance to demonstrate that the solute mass continues to register as the solute dissolves.

Chemical stability

'If an atom has been filled up or [is] all ready full up (of 8 outer electrons) it becomes stable and therefore it is unreactive. The atom will stay that way forever and not react or loose or gain any electrons.'
Comment of student commencing post-16 chemistry[26].

Stability (and the related notion of lability) are important ideas in chemistry. The noble gases were formerly called the inert gases (*ie* the 'not labile' gases), and the stability of noble gas electronic structure is much emphasised at upper secondary level.

The reactivity series of the metals, and the order of reactivity of the halogens are used as key principles for explaining why certain reactions do, or do not, occur. Of course such explanations can readily become tautologies (see Chapter 3) – so we might say that chlorine will displace iodine from its salts because chlorine is more reactive, when we only know that chlorine is more reactive because of our observations of this and other reactions. Nevertheless, stability and reactivity are key ideas in chemistry.

Research suggests that students may often have a limited appreciation of ideas about stability and reactivity. In particular, learners may readily come to adopt ideas about stable electronic structures (*ie* 'full shells' or 'octets'), but then to focus on this factor to the exclusion of other considerations.

Two of the classroom resources included in the companion volume are designed to help teachers explore their students ideas about 'stability' and chemical reactivity.

Students' ideas about chemical stability

The **Chemical stability** probes are designed to explore how students judge the relative stability of related chemical species. The original form of this probe – an updated version of which is provided as **Chemical stability (1)** – used the three species Na, Na^+ and Na^{7-}, and was provoked by the finding that some students considered that the Na^{7-} anion would be stable because it had an octet structure.[27]

RS•C

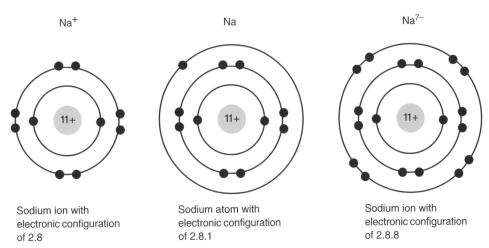

Na$^+$ — Sodium ion with electronic configuration of 2.8

Na — Sodium atom with electronic configuration of 2.8.1

Na^{7-} — Sodium ion with electronic configuration of 2.8.8

Figure 6.12 Representations of three species presented in Chemical stability 1 probe

The probe was first used with a group of sixteen post-16 students who had studied the topics of atomic structure and ionisation energies at college level. It was found that most (13/16) thought that the Na$^+$ ion would be more stable than the Na atom. Although the Na$^+$ species is commonly found as part of chemical systems, in isolation the neutral atom would be considered more stable. The cation would spontaneously attract a free electron, whereas energy is needed to ionise the atom.

Of more concern, over half the group (10/16) thought the neutral atom would be less stable than the Na^{7-} species – a highly charged metal anion. Half of this group also believed that this anion would be as stable as the Na$^+$ cation. The explanations given were usually in terms of the species with octets or full outer shells being more stable. (Of course Na^{7-} is actually 10 electrons short of a full outer shell, but this point was generally overlooked).

When the probe was given as an induction exercise to new students starting a post-16 chemistry course it was found that similar responses were obtained: 11/13 students thought that both ions would be more stable than the atom, and 9/13 thought that the anion was as stable as the cation.

When the probe was piloted for this project with two classes[28] of 15-16 year olds in a secondary school, it was found that nearly all of students (57/59, ie 97%) thought that the cation was more stable than the atom; and over four-fifth (51/59, ie 86%) though that the neutral atom was less stable than the Na^{7-} anion. Comments from students in this school included;

'[Na^{7-}] has a full outer shell of electrons therefore is more stable. [The Na atom] only has one outer electron and is less stable.'

[Na^{7-} and Na$^+$ are equally stable] 'because they both have full outer shells meaning they both do not need to lose or gain electrons'.

A few students even thought that the anion would be more stable than the cation – 'because both have a full outer shell but [Na^{7-}] has more of them so it is more stable'.

The probe has also been presented to student teachers. When 38 postgraduate trainee science teachers undertook the probe most (26/38, ie 68%) thought that the sodium cation was more stable than the atom, and a significant minority (15/38, ie 39%) believed that the anion was more stable than the neutral atom.[29] When a small group of Hungarian university students training to be chemistry teachers tried a version of the probe, a number (3/8) thought that the sodium cation was more stable than the atom, and one member of the cohort believed that the anion was more stable than the neutral atom.[30]

Alternative versions of the probe, **Chemical stability (2–7)**, have since been produced with other triads of species for several other elements. All of these are suitable for use in post-16 courses, and probes 1–4 are also suitable for students in the 14–16 age range who have a basic appreciation of atomic structure.

RS•C

When the probes[31] were piloted for the project they revealed that the focus on full electron shells was also found in other examples. The ions C^{4+} and C^{4-} were both judged more stable than a neutral carbon atom, and to be equally as stable as each other, 'because they both have full outer shells of electrons'. Similarly the beryllium atom was considered less stable than the Be^{6-} ion 'due to [Be^{6-}] having complete outer shell' which (according to one student) made it 'very hard to remove electrons' compared to the atom where 'electrons more easily removed'. Again the full shell criterion was seen by some students to have priority over any consideration of electrical neutrality. In one of the probes (probe 6) the Cl^{11-} ion was compared with the chloride ion and the neutral atom. Even such a highly charged species was thought to be more stable than the atom, as it had 'a full shell of outer electrons'.

1.8.8   2.7.8   2.8.7

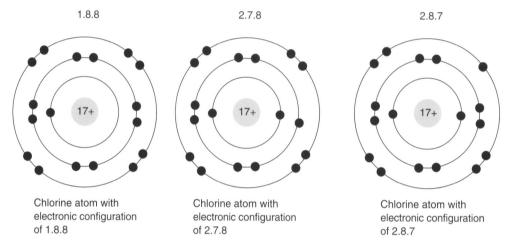

Chlorine atom with   Chlorine atom with   Chlorine atom with
electronic configuration   electronic configuration   electronic configuration
of 1.8.8   of 2.7.8   of 2.8.7

Figure 6.13 Representations of three chlorine atoms presented in the Chemical stability(5) probe

One of the probes (probe 5) takes a slightly different approach, asking students to compare a chlorine atom in the ground state (2.8.7) with two excited atoms of configurations 1.8.8 and 2.7.8 (see Figure 6.13).

One student suggested that the 2.8.7 configuration (ground state) atom was less stable than the (more excited, higher energy) option of a 1.8.8 configuration, as the atom with the 2.8.7 configuration 'requires 1 more electron to fill the outer shell unlike [the excited atom.]' A classmate who shared this belief that the ground state atom was less stable because the excited atom had a full [sic] outer shell, also thought that the atom with the 1.8.8 configuration would be more stable than the atom with the 2.7.8 configuration as the 1.8.8 configuration had both 'full outer shell and full second shell'.

In these responses we see that the criteria of a full outer shell can seem even more important to students than an atom having full inner shells. Clearly the 'full outer shell' notion is often used uncritically by students, without any deeper understanding of how it relates to other relevant factors.

Relating stability to reactivity

The **Chemical stability** probes (discussed above) are intended to allow teachers to see how strong their students adherence to the 'an octet structure is always more stable' rule. This alternative conception would seem to be quite widespread if the responses collected in the pilot are any indication. Whilst the suite of probes does provide teachers with a context in which to challenge this belief, it is an artificial comparison as it does not provide any context for judging the stability of the species.

Teachers clearly need to challenge the beliefs that lead to students arguing that Na^{7-} is a more stable species than Na^+. However, although it is interesting that many students consider the sodium cation to be more stable than the atom (despite the ionisation of the atom being an endothermic process), this is not such a surprising finding when the sodium ion, unlike the atom, is found as part of familiar stable systems.

RS•C

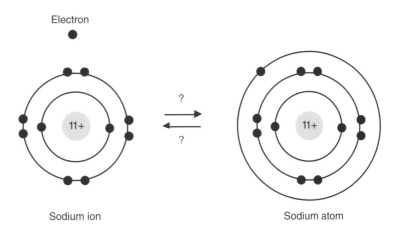

Figure 6.14 Two possible processes

The **Stability and reactivity** probe takes this comparison and places it in a context. Students are presented with a diagram (see Figure 6.14) which shows the possible processes of the atom being ionised to give a cation and an electron, and the cation joining with an electron to form an atom. The ionisation process requires an energy input, and will not occur spontaneously, whereas the positively charged ion and negative charged electron will spontaneously form an atom.

In the **Stability and reactivity** probe students are asked which of the two processes are likely to occur, along with questions about which of the ion and atom are (a) more stable, and (b) more reactive.

If students recognise that the electrical attraction will lead to the formation of the atom, but ionisation is not spontaneous, then the usual meanings of 'stable' and 'reactive' should lead to them judging the atom as more stable and the ion as the more reactive species in this context.

When the probe was piloted in schools and colleges it was found that students often held the alternative conception (that ion formation would be spontaneous) and for some students the response patterns were not so clear cut.

In one school over fifty 14–15 year old students responded to the probe. 43/54 students (ie 80%) thought the atom would emit an electron, and only one thought the atom and electron would combine. (Seven students did not think either process would occur, and three responded that they did not know the answer.) As one of the students explained: 'The sodium atom will emit an electron because it wants to have a full outer shell.' In this case there was some consistency, with most students considering the ion more stable and the atom more reactive.

This pattern of responses was also found among many post-16 students. One student in a group of 16–17 year olds in another school explained her responses thus:

'[The sodium ion is more stable than the sodium atom.] This is because the sodium ion has more of a full outer shell than the sodium atom.'

'[The sodium atom will emit an electron to become an ion.] To achieve stability by having a full outer shell.'

'[The sodium atom is more reactive than the sodium ion.] It has less of a full outer shell.'

However, many students do not produce responses which are consistent. For example, another student in the same group judged the ion to be both the more stable, and the more reactive of the species:

'[The sodium ion is more stable than the sodium atom.] The atom loses an electron to give it a full outer shell of electrons.'

'[The sodium ion is more reactive than the sodium atom.] Since the sodium ion has a positive charge it will readily attract electrons from another substance therefore reacting with that substance.'

RS•C

A small group of 13–15 year old High School students in Greece also responded to the probe. Most thought that the sodium ion was more stable than the atom (5/6, with 1/6 considering the atom more stable), and that the sodium atom was more reactive than the ion (4/6, with 1/6 believing the ion more reactive, and 1/6 unsure). All of the small group thought that the atom would spontaneously emit an electron.

When the small group of Hungarian university students training to be chemistry teachers attempted this probe they were divided over which process would spontaneously occur. Three of the eight in the group thought the cation and electron would join, an equal number expected the atom to spontaneously emit the electron, and the other students did not think either process would occur.

Perhaps part of the problem here is that students have understandable difficulty in considering the 'reactivity' of a chemical species in isolation: reactivity needs to be judged in a realistic chemical context. As discussed earlier in the chapter, students often tend to muddle ideas at the level of macroscopic phenomena and our molecular models (eg see Figure 6.11). Indeed, it has been suggested that to help students appreciate this distinction the word 'react' should be reserved for the macroscopic process, and the alternative term 'quantact' used to describe the molecular interactions that occur during a reaction (see Figure 6.15).[32,33]

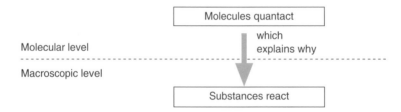

Figure 6.15 Quantaction and reaction

This terminology is not yet in common usage, but could help teachers maintain the distinction between the macroscopic level of observable phenomena, and the explanatory framework of the molecule model. If this suggestion is followed, then just as molecular level particles should not be said to melt during melting, or expand during thermal expansion, nor should they be said to react during reaction. The particles quantact, and this quantaction at the molecular level can be used to explain why reaction occurs at the macroscopic level.[34]

Notes and references for Chapter 6

1. P. Gailiunas, Is energy a thing? Some misleading aspects of scientific language, *School Science Review*, 1988, **69** (248), 587–590.

2. M. T. H. Chi, J. D. Slotta, & N. de Leeuw, From things to processes; a theory of conceptual change for learning science concepts, *Learning and Instruction*, 1994, **4**, 27–43.

3. A. Brook & R. Driver, in collaboration with D. Hind, Progression in Science: The Development of Pupils' Understanding of Physical Characteristics of Air across the age range 5-16 years, Leeds: Centre for Studies in Science and Mathematics Education, University of Leeds, 1989.

4. P. Johnson, Children's understanding of substances, part 1: recognising chemical change, *International Journal of Science Education*, 2000, **22** (7), 719–737.

5. K. S. Taber, Finding the optimum level of simplification: the case of teaching about heat and temperature, *Physics Education*, 2000, **35** (5), 320-325.

6. Of course this approach needs to be extended to ionic materials, where a compound will contain two or more types of ion rather than one type of molecule. Hydrated salts are usually considered as single substances despite being comprised of several types of ion and molecules of water. As was pointed out in Chapter 2, the basic concepts of chemistry are far from straightforward.

RS•C

7. The original student statement was that a 'Compound is one or more elements mixed together': K. S. Taber, Chlorine is an oxide, heat causes molecules to melt, and sodium reacts badly in chlorine: a survey of the background knowledge of one A level chemistry class, *School Science Review*, 1996, **78** (282), 39–48.

8. Research suggests that students tend to see science as largely about facts, and to have naive notions of how and why models are used in science (see notes 9 and 10). This situation seems to have persisted despite being raised as an issue twenty years ago (see notes 11 and 12).

9. R. Driver, J. Leach, R. Millar & P. Scott, *Young People's Images of Science*, Buckingham: Open University Press, 1996.

10. L. Grosslight, C. Unger, E. Jay & C. L. Smith, Understanding models and their use in science: conceptions of middle and high school students and experts, *Journal of Research in Science Teaching*, 1991, **28** (9), 799–822.

11. R. J. Osborne & J. K. Gilbert, The use of models in science teaching, *School Science Review*, 1980, **62** (218), 57–67.

12. N. J. Selley, The place of alternative models in school science, *School Science Review*, 1981, **63** (223), 252–259.

13. P. Johnson, Progression in children's understanding of a 'basic' particle theory: a longitudinal study, *International Journal of Science Education*, 1998, **20** (4), 393–412.

14. M. R. Abraham, E. Grzybowski, J. Renner, J. & E. Marek, Understanding and misunderstanding of eight grades of five chemistry concepts found in textbooks. *Journal of Research in Science Teaching*, 1992, **29** (2), 105–120.

15. T. Wightman, in collaboration with P. Green and P. Scott, The Construction of Meaning and Conceptual Change in Classroom Settings: *Case Studies on the Particulate Nature of Matter*, Leeds: Centre for Studies in Science and Mathematics Education, 1986, 217.

16. As one student explained 'it would seem a bit of an odd-ball' for one atom to have another atom's electron: K. S. Taber, An alternative conceptual framework from chemistry education, *International Journal of Science Education*, 1998, **20** (5), 597–608.

17. The student was undertaking an induction exercise at the start of an A level (UK) chemistry course. The two definitions which are discussed in the rest of this section had the same group as their source. For a discussion of suitable induction tasks at this level, and likely student responses, see K. S. Taber, Chlorine is an oxide, heat causes molecules to melt, and sodium reacts badly in chlorine: a survey of the background knowledge of one A level chemistry class, *School Science Review*, 1996, **78** (282), 39–48.

18. L. Renström, B. Andersson, and F. Marton, Students' conceptions of matter, *Journal of Educational Psychology*, 1990, **82** (3), 555–569.

19. During the piloting of the resource materials **Elements, compounds and mixtures** one of the teachers commented on how students in his class of 16 year olds 'wanted to see bonds as concrete structures, *ie* lines between cores'.

20. Figures 6.7 and 6.8 adapted from K. S. Taber, Molar and molecular conceptions of research into learning chemistry: towards a synthesis, 2000 – available via Education-line, at **http://www.leeds.ac.uk/educol/** (accessed September 2005).

21. P. Johnson, Children's understanding of changes of state involving the gas state, part 1: Boiling water and the particle theory, *International Journal of Science Education*, 1998, **20** (5), 567–583.

22. Qualifications and Curriculum Authority (QCA), *Standards at Key Stage 3 Science*, London: QCA, 2001.

RS•C

23. A. K. Griffiths, A critical analysis and synthesis of research on students' chemistry misconceptions, in H-J. Schmidt, *Proceedings of the 1994 International Symposium Problem Solving and Misconceptions in Chemistry and Physics*, ICASE [The International Council of Associations for Science Education] Publications, 1994, 70–99.

24. M. Slone & F. D. Bokhurst, Children's understanding of sugar water solutions, *International Journal of Science Education*, 1992, **14** (2), 221–235.

25. This possibility has been suggested by Prof. Robin Millar of York University, based on related research undertaken in the science education group at York.

26. K. S. Taber, An alternative conceptual framework from chemistry education, *International Journal of Science Education*, 1998, **20** (5), 597–608.

27. K. S. Taber, Case studies and generalisability – grounded theory and research in science education, *International Journal of Science Education*, 2000, **22** (5), 469–487.

28. The result from the top two ability sets are discussed. Although a third set also undertook the exercise it was found that their answers revealed a lack of the basis understanding of atomic structure which was needed to make any sense of the probe.

29. K. S. Taber, Trainee Science Teachers' Conceptions of Chemical Stability, 2000 – available via Education-line, at **http://www.leeds.ac.uk/educol/** (accessed September 2005).

30. This data was collected by Prof. Zoltán Tóth of the Chemical Methodology Group at the University of Debrecen, Hungary.

31. Five probes were piloted. Two more (probes 2 and 7) were added to the set during the process of revising materials for publication.

32. K. S. Taber, The campaign to stop molecules reacting, *Education in Chemistry*, 2001, **38** (1), 28.

33. Figure 6.15 adapted from K. S. Taber, Building the structural concepts of chemistry: some considerations from educational research, *Chemistry Education: Research and Practice in Europe*, 2001, **2** (2), 123–158, available at **http://www.uoi.gr/cerp/** or **http://www.rsc.org/Education/CERP/index.asp** (accessed September 2005).

34. The term 'interact' is too general, as it can apply to elastic collisions between molecules, and intermolecular bonding formation (both without reaction) as well as quantaction (leading to reaction).

RS•C

7.   Chemical structure

This chapter considers a key area of chemistry, that of chemical structure. It also reviews some of the research findings about learners' ideas about atomic structure and other chemical structures, and introduces some related classroom instruments included in the companion volume.

The structure of the atom

During their secondary education students are expected to learn about the structure of the atom, or – more correctly – to learn about a particular model of the structure of the atom (see the comments on models in Chapter 6).

The usual model of the structure of the atom met at this level consists of the nucleus at the centre of one or more shells of electrons. The electrons are usually shown (in two dimensional diagrams) as being placed on these circular shells (*eg*, see Figure 7.1).

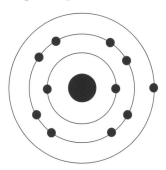

Figure 7.1 A typical representation of an atom

Although this model is perfectly appropriate at this level, those students who take their study of chemistry further (at post-16 level) will need to accept more detailed models (*eg* where electron positions are described by orbitals, which make up sub-shells). It is useful, therefore, for secondary teachers to emphasise that such a diagram only represents a model, and is one of several models that together help us understand matter at the atomic scale.

It is helpful if learners are familiar with a range of representations for molecules and other structures, as this will reinforce the modelling aspect. A mixture of different types of diagram have been deliberately used in preparing classroom materials for the companion volume.

The principles determining atomic structure

Clearly the atomic model is abstract, and a long way from learners' everyday experiences of the world. Students have never directly perceived individual molecules – except perhaps by smell, and that does not provide any insight to molecular structure. The terms 'proton' , 'neutron' and 'electron' are (initially) unfamiliar technical terms, and so need to be learnt by rote. As teachers are well aware, students may often confuse these labels while still mastering the basic model.

More significant than such errors are students' alternative ideas about how and why atoms, molecules and other chemical structures are formed and maintained. From the scientific viewpoint there are three main sets of principles involved:

■ the nucleus is held together by nuclear forces;

■ systems of nuclei and electrons (*ie* atoms, molecules etc) are held together by electrical forces; and

■ the tendency for these forces to minimise the energy of the systems is limited by quantization which restricts the allowed configurations.

RS•C

The nuclear interactions are usually taken for granted in chemistry, and only studied in physics. The importance of quantum restrictions is not usually referred to in the teaching of either chemistry or physics topics until post-16 level. Although the electrical nature of interactions may well be discussed, research suggests that students do not always appreciate the nature of the electrical forces involved. This means that atomic structure is normally taught without reference to two of the three scientific principles on which it is based, and that the one key principle which is considered may not be emphasised strongly enough.

In the absence of a sound physical basis for understanding chemical structures, it is not surprising that learners often develop their own alternative ideas.

Learners' ideas about the atomic nucleus

The term 'nucleus' itself may sound quite similar to 'neutron' and this may be a source of confusion. More significantly, students will be familiar with the use of 'nucleus' in biology and may sometimes – hard as it may seem to appreciate – confuse atoms and cells. (It is reported that a significant minority of students may consider atoms to be alive, perhaps viewing them as something like amoeba.[1])

In one sense such a comparison is impressive: cells are sometimes considered to be the 'building blocks' of organisms, and atoms are often said to be the 'building blocks' of matter (even though this simplistic view is problematic, see Chapters 6 and 10). The cell-nucleus–atomic-nucleus analogy can be significant. The cell nucleus is often described as a type of 'control centre' for the cell, and the atomic nucleus may be understood to be a control centre for the atom. (This may contribute to the way that some learners see the force between nuclei and electrons to be unidirectional – from the nucleus, acting on the electrons.)

Making comparisons between different ideas is an important part of developing new concepts (see Chapter 2), but learners need to be taught to look for the negative as well as the positive aspects of an analogy. An example of this – seeing the atom as like a tiny solar system – will be discussed below.

If students appreciated the major role of electrical forces in maintaining atomic and molecular structures, then they might be expected to commonly ask how the nuclei – containing several (and sometimes many) positive charges are held together. Secondary students will not normally have considered the nature of nuclear forces, and might well expect the nucleus to be forced apart by the repulsion between the protons. That few students seem to spontaneously think of this problem seems to reflect the way that atoms are not usually conceptualised in electrical terms. This is unfortunate, as students are left without an appropriate way of thinking about the nature of chemical stability (see Chapter 6) and chemical reactions (see Chapter 9).

One suggestion to explain nuclear stability, mooted by post-16 level students, was that the nucleus was held together because of some influence from the electrons.[2] One student, Annie,[3] made such a suggestion in three different interviews months apart. During the first year of her two year course she suggested 'forces from the outer ring [sic]' were 'pushing' the neutrons and protons together. In a later interview she suggested that '[be]cause the nucleus pulls in the electrons, so [I don't know] if the electron forces actually help bind the nucleus, in any way' . At the end of her course she commented that 'obviously the electrons … may sort of control what's actually happening in the nucleus. Sort of … holding the neutrons and the protons together'

Another student, Carol mused about why a nucleus would be stable:

'you would think that a nucleus wouldn't be there really because, it's all protons and they repel, 'cause they're the same charge ... but, there's another force, might be to do with electrons around the outside that holds it together … acting from outside.'

These comments reflect a common finding that students are often either ignorant of basic electrical principles, or at least do not transfer them from 'physics contexts', to apply them in the 'chemical context' of atoms and molecules. As one post-16 student taking college courses in both physics and chemistry explained:

RS•C

'I can't think about physics in chemistry, I have to think about chemical things in chemistry.'[4]

This 'compartmentalisation' of learning may well be partly responsible for some of the common alternative conceptions that students hold about atomic structure.

Learners' ideas about atomic structure

This lack of application of basic electrical ideas to the atom is reflected in the way students often conceptualise the way the electrons are held in position around the nucleus.

According to accepted scientific principles:

- all electrons in an atom are attracted to the nucleus;

- the force acting on an electron (due to the nucleus) depends upon the magnitude of the nuclear charge and the separation (the distance between the electron and the nucleus);

- the attractive force between an electron and the nucleus acts in both directions: both experience the same magnitude force; and

- each electron repels the others with a force which depends upon their separation.

To help simplify more complex atoms, we often introduce the idea of 'shielding' where inner shell electrons are considered to cancel the effect of an equivalent number of nuclear protons, so we can model the atom as a positive core charge and one shell of outer or valence electrons. This is only partially valid, as the 'electron shells' are not actually shells and interpenetrate – but it remains a useful approach. However, it is difficult for students to appreciate how the concept of shielding is supposed to work unless they accept the principles above.

It may seem that these points are the domain of physics rather than chemistry, yet these principles become quite important when students study chemistry at post-16 level, and are expected to explain such phenomena as patterns in ionisation energies. It is therefore significant that considerable numbers of students may well have alternative ideas about aspects of these interactions.

Interviews with post-16 students studying chemistry revealed the following alternative conceptions:

- the nucleus is not attracted by the electrons;

- the nucleus attracts an electron more than the electron attracts the nucleus;

- the protons in the nucleus attract one electron each; and

- the electrons repel the nucleus.

A classroom probe designed to elicit these, and related ideas, from post-16 level students is included in the companion volume. The **Ionisation energy** probe was originally used with 110 students in one college,[5] and was then piloted for the present publication with responses from 334 students in 17 different schools and colleges.[6]

It was found that just over a quarter of this sample of post-16 level students agreed with each of the statements that 'Each proton in the nucleus attracts one electron' and 'The nucleus is not attracted to the electrons.' Where students did think that the nucleus would be attracted to the electrons, they tended to agree most with statements suggesting the force on the electron would be larger than the force on the nucleus.

Half of these students agreed that 'Electrons do not fall into the nucleus as the force attracting the electrons towards the nucleus is balanced by the force repelling the nucleus from the electrons.'[7]

RS•C

Conservation of force: a common alternative conception

The interview study with post-16 students referred to above also found that students commonly held an alternative view of the way the nucleus held electrons in the atom. According to science the force between the nucleus and an electron depends upon the size of their charges and their separation.

Yet according to the alternative view the force was due entirely to the nucleus, and the size of the nuclear charge determined the total amount of force the nucleus could 'provide'. Force was seen as originating out of the nucleus towards the electrons. As the total nuclear force was fixed it would be shared by the electrons in the atom. This common alternative conception is known as the 'conservation of force' conception.[8]

Although this idea is incorrect it can be strongly held by students. In part this might relate to notions (referred to above) of the nucleus being the atom's control centre. However, perhaps the main reason for this was that the principle could be used to make correct predictions:

■   the larger the nucleus the more strongly electrons are attracted (true: helium has a higher ionisation energy than hydrogen);

■   the less electrons the more strongly they are each attracted (true: when an atom is ionised the second electron becomes more difficult to remove).

Although the reasoning is not quite correct, the use of this principle is reinforced when it seems to work successfully.

Several items in the **Ionisation energy** diagnostic instrument relate to this alternative conception, and have been found to be accepted by most students in the sample from 17 institutions:

■   'If one electron was removed from the [sodium] atom the other electrons will each receive part of its attraction from the nucleus' – 55% of the sample agreed.

■   'The third ionisation energy [of sodium] is greater than the second as there are less electrons in the shell to share the attraction from the nucleus' – 57% of the sample agreed.

■   'After the [sodium] atom is ionised, it then requires more energy to remove a second electron because once the first electron is removed the remaining electrons receive an extra share of the attraction from the nucleus.' – 61% of the sample agreed.

Indeed, the clearest statement of this 'conservation of force' principle – 'The eleven protons in the [sodium] nucleus give rise to a certain amount of attractive force that is available to be shared between the electrons' – was judged to be true by 72% of these college level chemistry students, and was considered false by only 15%.

Learning by analogy – the example of the atom

In Chapter 4 the notion of learning impediments was discussed. Meaningful learning relies on the learner interpreting new information in the context of what they already know – so that it 'makes sense' to them. It was suggested that sometimes when students fail to learn the science that is presented to them, this may be due to understanding differently, when they relate the new material to alternative conceptions they already have. Much of the material in this publication is concerned with helping teachers explore students' alternative conceptions.

However, it was also suggested that sometimes students simply fail to make sense of teaching because they cannot relate what they are hearing and seeing to any existing knowledge. Where such 'null impediments' occur, teachers need to find ways to bridge between the new knowledge and what the learner does already know. Often the pre-requisite learning is in place, and the teacher simply needs to make the connections more explicit.

When the new ideas are too abstract to be directly related to existing ideas, teachers often call upon comparisons with other more familiar contexts. Atomic structure is clearly a topic which is highly abstract, as students are expected to learn about the internal structure of a conjectured entity which is

RS•C

much too small to be directly experienced. We saw in Chapter 6 that many students had difficulty enough making sense of the molecular model of matter – the interactions within an atom are a further step from their everyday experience.

Not surprisingly analogies are often used when teaching about the structure of the atom. Various fruit placed at the centre of a large room may be used to give some feel for the scale of the nucleus within the atom. (Alternatively references to balls in various sports venues could be used – see Chapter 10.) A common comparison that is made is that 'an atom is like a tiny solar system'. The relationship between the nucleus and electrons is here modelled on the sun and planets. It has been suggested that although this may be useful for giving students an image of the atom, it is an approach that can go wrong.[9]

The use of this teaching analogy relies upon a number of assumptions;

1. that an atom is in some ways like a solar system;

2. that the students are familiar enough with the solar system to make use of the comparison; and

3. that the students will recognise in which ways the atom is like a solar system, and in which ways it is not.

None of these points are straightforward. The 'planetary' model of the atom is only of limited use once students move into post-16 courses, when they will be expected to see atoms in terms of orbitals (rather than orbits) and 'clouds' of electron density. In any case students may well already be familiar with the image of the planetary atom as it is a cultural icon that they will have seen in many contexts. Students' understanding of the solar system may not be orthodox, and they may not be clear about which features are to be transferred onto the atom.

These criticisms are not intended to suggest that such analogies should not be used. However, research does suggest that currently teachers do not always help students understand which features of an analogy or model do or do not match the target.[10] The significance of highlighting similarities and differences for forming new concepts was discussed in Chapter 2, and practice in such comparisons may be very useful to students.

One of the classroom activities included in the companion volume, **An analogy for the atom**, provides teachers with a chance to explore their students' use of this teaching analogy. This activity has two parts. The first worksheet, **The atom and the solar system**, provides a probe for eliciting students' ideas about the forces at work in these two systems.

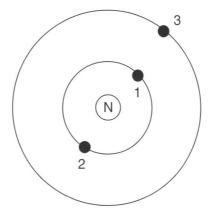

Figure 7.2 A simple representation of an atom

When this was piloted for the project, it was found that students often held alternative ideas about both the atom and the solar system. For example, one student in a class of 14–15 year olds who had studied atomic structure reported that the type of force attracting the electron towards the nucleus was a 'pull' force. An electron further from the nucleus (electron 3 in Figure 7.2) would be attracted by a stronger force as 'it is further away and therefore it will need a stronger force to draw the

RS•C

electron towards the centre'. He thought that there was no force acting on the nucleus due to an electron (as 'the electron is drawn to the nucleus instead of the nucleus [being] attracted [to] the electron'). He also thought there would be no force between the electrons as 'they are fixed on an axis and they have to have a fixed distance away from each other'. This student seemed to have quite firm ideas about atoms, albeit ideas at odds with the scientific model.

Other students in the same class suggested that the force attracting the electrons toward the nucleus was 'gravity', 'magnetic' or 'pole force'. Some classmates agreed that the nucleus would not be subjected to a force as 'the nucleus is the only thing that can apply a force'. Some students thought that there was a force attracting the nucleus towards the electrons, although it would be smaller (than the force attracting the electrons to the nucleus) as the 'nucleus is bigger + [has] bigger mass so bigger force'. Some students thought the electrons could not be interacting with each other, as they were interacting with the nucleus, or there was 'no relationship between them', whilst others acknowledged a gravitational 'reaction'. Some students did know electrons would repel each other, and one suggested 'they repel each other around the nucleus'.

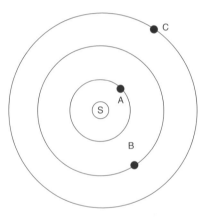

Figure 7.3 A simple representation of a solar system

Most of the students in this class recognised the role of gravity in attracting planets to the sun, although the 'pole force' also put in an appearance. Most of the group did not think the planets exerted a force on the sun, and about half thought there were no forces between planets. Some of the comments reflect the answers to the questions about the atom. The force acting on the planet with the largest orbit had to be greatest as 'it's further away so the force to stay with [the sun] is much bigger', 'the planets are attracted to the sun, not the other way round' and that 'planets [are] only attracted to the sun' and not each other.

Although the similarities in response suggested that learners might well see similarities between the two systems, the high proportion of alternative notions suggests that using the comparison as a teaching analogy could simply transfer incorrect ideas about one system to the other.

The second part of the exercise, **Comparing the atom with the solar system**, asks students to list the similarities and differences between the atomic system (Figure 7.2) and the solar system (Figure 7.3). To some extent this activity is scaffolded (see Chapter 5), as the questions asked on the **The atom and the solar system** worksheet provide some cues for suitable comparisons. The first worksheet can be seen as organising the students' existing knowledge to prepare them for the later task. It acts as a scaffolding PLANK, providing a conceptual platform for developing new knowledge.

When this exercise was piloted for the project it was found that some students found it very difficult to suggest more than a couple of similarities or differences (and many did not make the 'obvious' point about the atom being a good deal smaller than a solar system). This may suggest that this is a skill which needs more explicit practice (see the exercises on **Chemical comparisons** discussed in

RS•C

Chapter 2). The class of 14–15 year olds which produced the responses discussed above did have a fair attempt at spotting similarities and differences. Some good suggestions were made for both the similarities, and the differences:

Similarities:

■ Both the atom and the solar system have centres that attract the surrounding planets or electrons

■ Both have forces involved

■ They both rotate around a centre point

Differences:

■ More than one thing on the ring in atoms

■ Planets have no charge but electrons are negatively charged

■ Electrons have virtually no mass and planets have a large mass

■ The solar system is a lot bigger

■ Different forces

■ Planets can be seen with the naked eye, electrons can't

■ The planets rotate around

However, unsurprisingly in view of the alternative notions revealed in the first part of the activity, some of the points of similarity and difference suggested were:

Similarities:

■ They are both [electrons, planets] held in orbit by gravity

■ They [electrons, planets] are not attracted to each other

■ They both have energy sources in the centre

■ Neither the nucleus or the sun are attracted to the planets or electrons

Differences:

■ The planets move around [implying that the electrons do not]

■ The rings are closer together around the sun

■ The force attracting the particles is pull but the force attracting the planets is gravity [implying gravitational force is something other than a pull]

■ There is force between [the planets – implying no force between electrons]

Learners' ideas about orbitals

Post-16 level students are often expect to move beyond ideas about electron shells, to learn something about electronic orbitals. Some observers feel that orbital ideas are unhelpful prior to university level study. It has been suggested that the notion of 'electron pair domains' is simpler, and sufficient for school and college level study.[11] However, examination stipulations may require learners to tackle orbital concepts.

This has found to be a topic that students often find difficult.[12] This should not be surprising because orbital ideas are highly abstract, and so students may find difficulty making sense of them in terms of existing knowledge (*ie* there may be a 'null learning impediment' – see Chapter 4). Where students have a naive appreciation of the roles of models in science (see Chapter 6), they may well be committed to the idea of electron shells, and this existing school science knowledge may interfere with the intended learning (*ie* there may be a 'pedagogic learning block' – see Chapter 4).

RS•C

Students often come to this topic with an image of the electron shell (usually represented as a circle around the nucleus) as a kind of electron orbit, and so the term 'orbital' may initially be acquired as an alternative term with much the same meaning.[13] When the idea of sub-shells is introduced, this may again become confused with shells, orbits and orbitals. Further confusion is likely as other related concepts are introduced. So orbitals may be confused with energy levels, and conventional diagrams representing orbital 'probability envelopes' may be read as showing orbital boundaries.

Until students have managed to differentiate between some of these related, but distinct, concepts, the conventional labelling for atomic orbitals (1s, 2s, 2p, 3s, ...) is meaningless, and unlikely to be mastered.

Such abstract ideas are demanding even for able students, and time, reinforcement and practice are needed if learners are to show (and maintain) a good understanding of the orbital topic. Unfortunately, the pressures of covering the course material often mean that before this can occur, students have already been introduced to further complications: hydbridization and molecular orbitals. The latter may be σ or π, and may be bonding or anti-bonding. The folly of expecting some students to make sense of these ideas after a limited exposure is reflected in the way one student in New Zealand defined anti-bonding orbitals as 'silly things' that just stuck out![14]

One particular problem that some students have is recognising that rehybridization of atomic orbitals (a formal mathematical process used in trying to understand nature, not a process in nature itself!) gives a set of atomic orbitals – some of which will no longer be present once a molecule is formed. It may help us to understand the bonding to think of a carbon atomic system undergoing rehybridisation to give sp^3 atomic orbitals suitable for overlap. There are no electrons in a methane molecule, however, best described as being in sp^3 hybrid orbitals – those electrons are now in σ-molecular orbitals.[15]

In a molecule as simple as oxygen (O_2) there will be orbitals that are effectively atomic orbitals unchanged from those in a ground state atom, orbitals which are effectively hybridised atomic orbitals, and molecular orbitals which are unlike any of the precursor atomic orbitals. The level of treatment expected in post-16 chemistry is likely to consider how rehybridisation allows two partially occupied orbitals on each atom to overlap to form molecular orbitals. The overlap of sp^2 hybrids allows the formation of the σ-bond, and the overlap of unhybridized p orbitals allows the formation of the π-bond.

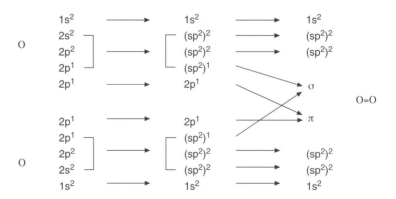

Figure 7.4 A scheme to show the orbitals in a simple molecule

Figure 7.4 shows the (hypothetical) process of moving from ground state atomic orbitals on two oxygen atoms, through rehybridised atomic orbitals, to the orbitals expected in the double bonded molecule. Even in such a relatively straightforward example as a diatomic molecule of an element, it is little surprise that such a scheme proves confusing to many students. It is therefore rather ironic that oxygen is found to be paramagnetic – and must therefore have unpaired electrons in its ground state. The model used at post-16 level does not even predict the correct electronic structure in this case.[16]

At this level radicals are considered to always be highly reactive species, the diradical nature of oxygen[17] is not normally discussed, and students assume that electrons are paired up in the oxygen molecule.

Depicting molecular structure

In this publication a deliberate attempt has been made to use a variety of ways of representing molecules and other structures. (For example, in Chapter 5, when suggestions are made about introducing hydrogen bonding through the structure of the water molecules, the example worksheet included three distinct representation of the molecules.) This is because all of our diagrams are limited ways of representing various aspects of our mental models of molecules. The importance of models in chemistry, and the limited way in which they are appreciated by students, was considered in Chapter 6. Whenever we chose to use a particular type of diagram we are (consciously or not) emphasising certain aspects of our models of molecules. Often some aspects of the diagram are irrelevant or even unhelpful, and it may not always be clear to students which aspects of a diagram (or a solid model or computer animated model) they are meant to be attending to, and which aspects are less relevant in a particular context. It has been found that students' ability to solve chemical problems is often related to their ability to interpret different representations of chemical systems.[18]

It is suggested that teachers should both use a variety of diagrams and other models, and make a habit of reiterating which aspects of the model are and are not significant in a particular context. Reference was made in Chapter 2 to some research where students were asked to make discriminations between different chemical species (molecules, ions, atoms etc) represented by pictures from textbooks. It was found that some of the students largely used criteria which were based purely on the way the species were drawn, rather than their chemical attributes. That is, some learners 'seemed to discriminate between [chemical species] on the basis of the way they were represented, rather than what was represented'.[19] Where some students focused on features of the chemical species themselves, others would comment upon 'the different conventions used in chemistry textbook diagrams to represent various aspects of the species drawn' such as whether electrons were shown as 'e' or '•'.

Noticing such different conventions does not always equate to considering them significant, but does remind us that aspects of diagrams that have effectively become invisible to 'experts' due to familiarity may draw the attention of relative 'novices'. Such aspects may act as distractions, and take up some of the limited 'slots' in the students' 'mental scratch pad' (see Chapter 5). This is another example of why it is important for the teacher to learn to see the material presented at the 'resolution' available to the learner.

Consider the following diagrams:

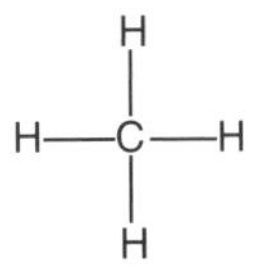

Figure 7.5 A representation of a methane molecule

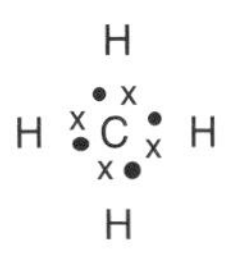

Figure 7.6 A second representation of a methane molecule

RS•C

Figure 7.7 A third representation of a methane molecule

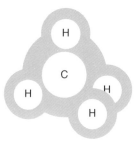

Figure 7.8 A fourth representation of a methane molecule

The four diagrams in Figures 7.5–7.8 are different ways of representing a methane molecule. In Figure 7.5 the bonds are shown as lines. This is a type of diagram that students can easily learn to draw, but it may not always be the most appropriate diagram to use. Younger students are known to sometimes think of chemical bonds as being physical connections between atoms – and to think of bonds as (and not just analogous to) sticks, springs or adhesive.[20,21] This type of diagram could reinforce such a view.

Figure 7.6 actually could help to avoid this as it represents the bonding electrons. The dot and cross formalism is very common and is meant to help students with their electron-accounting by showing where the electrons (are conjectured to have) originated. However, this type of figure may encourage common alternative conceptions that electrons from different atoms are different and are only (or are more) attracted to their own nucleus, and will always return to the original atom when the bond breaks. Another source of confusion is that the same dot and cross symbols may be used in a distinct convention to indicate the two different spin states of electrons rather than their atomic origins.[22]

Figures 7.5 and 7.6 may both give the impression that molecules are flat, whereas Figures 7.7 and 7.8 should avoid this interpretation by representing the three dimensional shape of the molecule. Figure 7.7 again uses lines for bonds, where Figure 7.8 gives an impression of the electron cloud. This type of diagram is a little more sophisticated, but gives a better impression of the way bonding electrons are part of molecules (rather than belonging to specific atoms) and the way molecules have quite 'fuzzy edges' rather than being like marbles with a distinct edge.

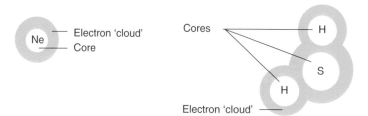

Figure 7.9 Representing molecules with electron clouds

This type of representation (see Figure 7.9) has been used quite a lot in this publication, and is used in some of the materials intended for 11–14 year old students. It might be objected that the idea of an electron cloud is far too abstract for students at this level, but all of these representations are abstract, and pictures with electron 'clouds' need be no more problematic than diagrams showing concentric electron 'shells', or representing bonds as lines or as dot and cross pairs. While students just beginning to learn about particle theory are not ready to fully appreciate what is meant by an atomic core or an electron cloud, these types of diagrams did not seem to be problematic for the students. It

RS•C

could be argued that showing electron clouds as shading is actually less abstract than the use of 'C' to represent the carbon nucleus and the K electron shell in Figure 7.5 – something we routinely expect pupils to accept.

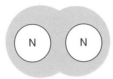

Figure 7.10 A representation of a nitrogen molecule

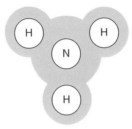

Figure 7.11 A representation of the pyramidal ammonia molecule

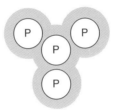

Figure 7.12 A representation of the tetrahedral phosphorus molecule

However, I am certainly not suggesting that this type of diagram is a panacea. Figure 7.10 does not clearly indicate the multiple nature of the nitrogen bond, and Figures 7.11 and 7.12 do not discriminate between the two different arrangements present. The type of diagram we select to use at any particular time should reflect the specific molecular features we are trying to emphasise.

Figure 7.13 A representation of a chlorine molecule

A diagram which is effective at making some points, may be less useful on other occasions. We also need to be very careful in checking what students read into diagrams. Figure 7.13 shows a common type of diagram used in chemistry books. The overlap between atomic shells is meant to indicate the covalent bond, with the bonding electron pair within the overlap. One college student suggested that the bonding electrons in such a diagram were more restricted than the non-bonding (lone pair) electrons as they could only move within the area of overlap, whereas the other electrons shown could move throughout the shell.[23]

As this student started to learn a more sophisticated model of the electronic structures of atoms and molecules, she over-interpreted the more simplistic model of a molecule so familiar from her school science. This is not an argument for not using such diagrams, but as with teaching analogies (see earlier in this chapter), and all other forms of teaching models (see Chapter 6), we need to help students see which features are important, and which are just conventions of graphic artists. Selecting different types of diagrams of the same species, to emphasis different teaching points, can help our

RS•C

students realise which aspects of these representations are actually meant to reflect the abstract features of the species chemists represent.

Learning about lattice structures

It has been found that when students commence university study of chemistry they may often have a very limited understanding of lattice structures.[24] This is unfortunate as a key objective of learning about chemical structure is to be able to explain the properties of materials – hardness, electrical conductivity, cleavage planes in salts, malleability etc – and this requires an understanding of both chemical bonding (see the next chapter) and larger scale structure. One common problem is that, having learnt about molecules, students often assume that all materials are molecular.[25]

In ionic materials, the formula may seem to imply molecules. So NaCl is often thought to comprise of 'NaCl' molecules arranged in a lattice (see Chapter 8). Students who have learnt to use the idea of valency or combining power (H=1, N=3, C=4 etc) to work out the stoichiometry of covalent molecules (NH_3, CH_4 etc) will often extend this idea to the ionic case using electrovalencies (Mg=+2, F=−1, so magnesium fluoride will be thought to comprise of $Mg^{2+}(F^-)_2$ molecules). The diagrams used in many student textbooks are likely to encourage this false interpretation (see Chapter 10).

Similarly, in metals, the student may assume that the electronic configuration of the metal atom determines how many other metal atoms it bonds with, and therefore how many metal atoms are present in the 'molecules'.

In the case of substances with covalent bonding students may become confused as they need to distinguish between molecular materials bound by van der Waals forces, and giant covalent structures. Students sometimes suggest that covalent bonds are relatively weak because (for example) sulfur is readily melted and sugar easily dissolved. They often fail to realise that the intramolecular bonds may be unaffected in such processes.

Although there are only a limited number of familiar substances which have giant covalent lattices, this is an important type of structure. Learners may consider this type of material to contain discrete molecules with strong intermolecular forces, or may even consider solid carbon to comprise of discrete atoms, something that may in part derive from the 'molecular' formula of carbon being commonly given as 'C' and taken to imply 'C_1', when 'C_∞' might be more appropriate.

One of the classroom resources included in the companion volume, **The melting temperature of carbon**, is designed to challenge this alternative conception, and reinforce the relationship between melting temperature and structure. This study activity has two parts: **Predicting the melting temperature of carbon** and **Explaining the melting temperature of carbon**.

Predicting the melting temperature of carbon presents students with a table of the melting temperatures and relative molecular masses for a number of non-metallic elements (reproduced as Table 7.1). Although carbon is included in the table, its melting temperature is omitted. Students are asked to spot the pattern in the data provided (*ie* that melting temperature increases with increasing relative molecular mass) and to predict the melting temperature of carbon.

Element	M_r	Melting temperature/K
Helium, He	4.0	4
Carbon, C	12.0	
Neon, Ne	20.2	25
Fluorine, F_2	38.0	53
Chlorine, Cl_2	71.0	172
Bromine, Br_2	159.8	266
Iodine, I_2	253.8	387
Sulfur, S_8	256.8	392

Table 7.1 Comparing molecular mass and melting temperature

When the exercise was piloted a typical value predicted by many students was about 14 K. If students recall that carbon (graphite or diamond) has a high melting temperature, this will already present them with a puzzle to be solved.

Some students will not be aware of the puzzle, either not having learnt, or not bringing to mind, the fact that carbon has a high melting temperature. They are told that carbon has a melting temperature of 3823 K at the beginning of **Explaining the melting temperature of carbon**, so that they will all be aware of the anomaly.

The students are then asked to consider diagrams showing discrete molecules of neon, chlorine and sulfur alongside a similar style of diagram showing part of a 2-dimensional representation of a diamond lattice. The first three diagrams show discrete species (*eg* see Figure 7.13). The figure for carbon is clearly different, showing only part of a more extensive structure (see Figure 7.14).

Figure 7.14 Part of the giant covalent structure of carbon

The questions in this part of the exercise ask students if the diagrams help them explain why carbon has a high melting temperature.

A school student in a class of 14–15 year olds predicted that carbon would melt at about 13 K as the 'molecular mass of carbon is roughly half that of neon so the melting temperature is roughly half', and then explained the high melting temperature from the diagram, as 'every single carbon atom is bonded to 4 others carbon atoms meaning it takes a lot of energy to break the bonds'.

One post-16 student who predicted that the melting temperature of carbon would be 14 K because 'its molecular structure is a single atom...', later – after considering the diagrams – explained that 'you can see that the atoms are densely packed together with one carbon atom attached covalently to four [other] carbon atoms...so it means a lot of atoms bonded together'. Another student in the same post-16 class who predicted a melting temperature of 13.5 K explained that 'the C atoms carry on continuously bonding to other C atoms and the molecule does not stop' and that it was 'difficult to break the bonds and therefore difficult to melt'.

RS•C

Challenging students' expectations

This exercise is certainly contrived, being designed to encourage students to make an incorrect prediction, which then needs to be explained away. It also leads them to the type of explanation needed through the diagrams provided. For those students who are well aware of the high melting temperature of carbon, and the reasons for it, it will seem an artificial task, but provides useful practice in a number of areas (calculating molecular masses; finding patterns in data; explaining properties in terms of structure).

For those students who do not make a clear distinction between substances with covalent bonding which are simple molecular, and those which have giant lattices, the task presents a genuine anomaly.

Evidence from historical studies suggest that it is the anomalies between theoretical predictions and actual observations which often act as the incentive for conceptual revolutions. In a similar way, conceptual change in learners is believed to be encouraged by making them see how their ideas are inconsistent with actual observations (see Chapter 10). It is said that conceptual change may be brought about by 'dis-equilibrium' or 'cognitive dissonance' (where new information does not match existing ideas). As rational beings, we try to find ways to make our ideas more coherent.[26]

In laboratory situations phenomena are often complex, and in practice students often 'see' what they expect. By providing a data based exercise **The melting temperature of carbon** provides a definite anomaly that needs to be explained.

Notes and references for Chapter 7

1. A. K. Griffiths & K. R. Preston, Grade-12 students' misconceptions relating to fundamental characteristics of atoms and molecules, *Journal of Research in Science Teaching*, 1992, **29** (6), 611–628.

2. K. S. Taber, Understanding Chemical Bonding – the development of A level students' understanding of the concept of chemical bonding, Ph.D. thesis, University of Surrey, 1997.

3. An assumed name. As is standard practice, the names given to individual students discussed in the text have been changed to protect anonymity.

4. K. S. Taber, The sharing-out of nuclear attraction: or I can't think about physics in Chemistry, *International Journal of Science Education*, 1998, **20** (8), 1001–1014.

5. K. S. Taber, Ideas about ionisation energy: a diagnostic instrument, *School Science Review*, 1999, **81** (295), 97–104.

6. Details are available at the project website: **http://uk.groups.yahoo.com/group/challenging-chemical-misconceptions** (accessed September 2005).

7. This statement contains two major errors. Firstly it is implied that forces acting on two separate bodies can somehow cancel each other. Secondly, the positive nucleus is said to be repelled by the negative electrons.

8. The label 'conservation of force' refers to how a set amount of force is considered to 'emanate' from a nucleus of a particular charge regardless of the number and arrangement of electrons it is bound to. Of course force is not conserved in this way, although both charge and energy are conserved. There is some evidence from interviews that students do sometimes confuse the charge, with the force acting due to the charge; or the force with the work done by the force. (See note 2).

9. K. S. Taber, When the analogy breaks down: modelling the atom on the solar system, *Physics Education*, 2001, **36** (3), 222-226.

10. S. Sizmur & J. Ashby, *Introducing Scientific Concepts to Children*, Slough: National Foundation for Educational Research, 1997.

RS•C

11. R. J. Gillespie, Bonding without orbitals, *Education in Chemistry*, 1996, **33** (4), 103–106.

12. K. S. Taber, Building the structural concepts of chemistry: some considerations from educational research, *Chemistry Education: Research and Practice in Europe*, 2001, **2** (2), 123–158, available at **http://www.uoi.gr/cerp/** or **http://www.rsc.org/Education/CERP/index.asp** (accessed September 2005).

13. This could be another example of what Prof. Hans-Jürgen Schmidt calls a label acting as a hidden persuader – see Chapter 1.

14. R. K. Coll and N. Taylor, *Alternative conceptions of chemical bonding for upper secondary and tertiary students*, *Research in Science and Technological Education* **19 (2)**, 171–191.

15. The habit of describing electrons in molecules as though they are still in atomic orbitals may be related to the tendency to consider bonding electrons as still belonging to specific atoms.

16. The molecular structure of oxygen is better understood in terms of the nitrogen molecule, formed by overlap of sp hybrids and unhybridized p orbitals in two planes. In nitrogen this gives a triple bond (with a s-component and two p-components) and one non-bonding pair of electrons in sp-hybrids on each nitrogen atomic centre. In oxygen there are two more electrons to be accommodated, which are placed in degenerate p-anti-bonding orbitals. Although there is a triple bond structure, the overall bond order is 2 due to the anti-bonding orbitals. See N. W. Alcock, *Bonding and Structure: Structural Principles in Inorganic and Organic Chemistry*, Chichester, Ellis Horwood Limited, 1990.

17. J. Barrett, *Understanding Inorganic Chemistry: the Underlying Physical Principles*, Chichester, Ellis Horwood Limited, 1991.

18. G. M. Bodner & D. S. Domin, The role of representations in problem solving in chemistry, paper presented to the *New initiatives in chemical education* on-line symposium, available at **http://www.rsc.org/pdf/uchemed/papers/2000/41_bodner.pdf** (accessed September 2005)

19. K. S. Taber, Can Kelly's triads be used to elicit aspects of chemistry students' conceptual frameworks?, 1994, available via Education-line, at **http://www.leeds.ac.uk/educol/** (accessed September 2005).

20. T. Wightman, P. Green & P. Scott, T*he Construction of Meaning and Conceptual Change in Classroom Settings: Case Studies on the Particulate Nature of Matter*, Leeds: Centre for Studies in Science and Mathematics Education, 1986.

21. A. G. Harrison & D. F. Treagust, Learning about atoms, molecules, and chemical bonds: a case study of multiple-model use in grade 11 chemistry, S*cience Education*, 2000, **84**, 352–381.

22. A. J. Catterall, Electron structure in molecules: a new approach, *School Science Review*, 1967, **48** (166), 798–802.

23. K. S. Taber, Building the structural concepts of chemistry: some considerations from educational research, *Chemistry Education: Research and Practice in Europe*, 2001, **2** (2), 123–158, available at **http://www.uoi.gr/cerp/** and **http://www.rsc.org/Educaton/CERP/index.asp** (accessed September 2005).

24. D. Cros, R. Amouroux, M. Chastrette, M. Fayol, J. Leber & M. Maurin, Conceptions of first year university students of the constitution of matter and the notions of acids and bases, *European Journal of Science Education*, 1986, **8** (3), 305–313.

25. P. G. Nelson, To be a molecule, or not to be?, *Education in Chemistry*, 1996, **33** (5), 129–130.

26. Whilst it is undoubtedly true that we are driven to make sense of the world, and that coherence is an important criterion, we are also capable of holding apparently inconsistent views. Although being rational has its uses, we live in a world where information may be partial, incorrect and incoherent: so we have evolved brains that can cope with this muddled input. One consequence is the ability to hold manifold conceptions, or multiple conceptual frameworks, for the same topic – see Chapter 8.

RS•C

RS•C

8. Chemical bonding

Chemical bonding is one of the key concept areas in the subject, and is also an area where learners are known to commonly acquire alternative conceptions. Some of these alternative ideas are considered in this chapter, with suggestions for improving the teaching of the topic. Some classroom materials included in the companion volume for eliciting and challenging students' ideas are introduced.

The full shells explanatory principle

Students are found to commonly use the octet rule – a useful heuristic for identifying stable chemical species – as the basis of a principle to explain chemical reactions (see Chapter 9) and chemical bonding.[1] According to this 'full shells explanatory principle' bonding occurs 'in order [for atoms] to try to achieve a stable structure *ie* 8 electrons in the outer shell of the atom'.

Students relate the 'sharing' of electrons in covalent bonds to the full shells explanatory principle, so that 'the electrons are shared to create a full outer shell', and the 'covalent bond is the sharing of electrons to complete full valency shells'. As one student wrote in a test paper;

'A covalent bond is one in which two atoms join together by the sharing of electrons. Each of the atoms achieves noble gas configuration in the process of covalent bonding.'

Ionic bonding is similarly understood as 'where you, donate, or gain electrons, to form a completed outer shell'. The full shells explanatory principle may also be invoked in students' explanations of metallic bonding, so that one 18 year old student described how 'metals haven't got full outer shells, then by electrons moving around, they're getting, a full outer shell, but then they're sort of losing it, but then like the next one along will be receiving a full outer shell'. Another post-16 student conceptualised metallic bonding as being 'formed by the one, two or three valent shell electrons being donated to lattice so a noble gas configuration is achieved'.

The full shells explanatory principle is inherently anthropomorphic, as no physical force is invoked to explain why systems should evolve toward certain electronic configurations. Rather, it is assumed that this was what atoms 'want', and so they act accordingly: 'in all cases what an atom is trying to do is become stable'. For example, students may suggest that dimers of aluminium chloride form so that the aluminium atoms would 'think' that they have the octets they need to be stable.

Given that the starting point for many students' thinking about bonding is the atoms' perceived need to achieve a full shell, it is not surprising that often:

■ students see chemical bonding and forces within chemical structures as largely unrelated; and

■ students limit their category of chemical bond to types of interactions that can be readily conceptualised in terms of the full shells explanatory principle.

It would be an exaggeration to claim that all aspects of students' alternative ideas about chemical bonding derive from this invalid extension of the octet rule. Nevertheless, the full shells explanatory principle can be seen to be extremely influential in guiding the thinking of many students.

When is a chemical bond not a force?

We could define a chemical bond as that which holds the parts of a chemical structure together. Of course, the problems of working from definitions were discussed in Chapter 2, and as with so many other concepts in chemistry the experts (*ie* chemists and chemistry teachers) can feel they know what they think a chemical bond is, without necessarily being able to provide a rigorous definition.

However, in view of the range of alternative conceptions, some quite tenacious, that have been uncovered in this topic, it is sensible to start this chapter by exploring what a chemical bond is. I

RS•C

would suggest it is a force which holds chemical species together.[2] This force can usually be considered as an electrical interaction. The source of this interaction is the nature of chemical species themselves – composed of positively charged nuclei and negatively charged electrons.

It will be noted that this description would appear to be just as applicable to individual atoms or ions as to molecules and lattices. A sodium ion is held together by such electrical interactions, just as a hydrogen molecule, or an ice crystal is. By convention we do not usually refer to the interactions within a mononuclear species (a single ion or atom) as chemical bonding, but perhaps this is unfortunate as research shows that students commonly see the chemical bond as being unrelated to the electrical forces within a mono-nuclear species. Students often do not count 'intermolecular' interactions as chemical bonding either.

If there were no quantum restrictions on where electrons could be located in chemical systems, then teaching about and studying chemical bonding would be much simpler – we would not have different categories of chemical bond such as covalent and hydrogen bonding. (However, all the matter in the universe would collapse into 'neutron stars', so there would a high price to pay for this simplicity.) These quantum restrictions limit the number of electrons in particular 'shells' around nuclei, and lead to the 'complications' of, *inter alia*, electronic configurations, valency ('combining power'), electronegativity, and patterns in ionisation energies.

Unfortunately this complexity also commonly leads us to present the topic of chemical bonding to students in such a way that they do not appreciate the underlying electrical nature common to the different types of bonding.

Indeed a teacher might sometimes refer to bonding as a 'chemical bond', and other occasions refer to it as 'an interaction' or talk of, say, 'the force' holding molecules together. Yet different students may well have their own idiosyncratic ways of using these various terms that teachers are not aware of.

For example, one of the classroom resources provided in the companion volume is a probe called **Interactions** which asks students to describe and explain the interactions in a range of chemical species. It also asks students whether the interactions depicted would be classed as 'attraction', 'force', 'bonding' and 'chemical bond'. This may seem a strange and even pointless question, but interviews with college students have suggested that learners may use these terms in different ways – that is, differently to the teacher, and differently from one another.

Figure 8.1 Not a chemical bond?

When this probe was undertaken by a class of 16–17 year olds similar responses were made. One student described the interaction shown in Figure 8.1 as 'covalent', and reported that 'the two hydrogens are attracted to each other, so they are bonded together'. This particular student classed this interaction as 'bonding', but not as a 'chemical bond': but did think the interaction was both an 'attraction' and a 'force'. A classmate who agreed that this interaction was 'bonding' was unsure if it was also a 'chemical bond', but did not think it was either an 'attraction' or a 'force'. Another student in the class thought that this 'covalent bonding' was both 'bonding' and a 'chemical bond', however this student recognised the interaction as an 'attraction', but not as a 'force'.

RS•C

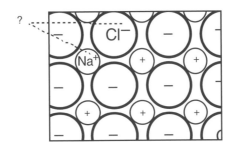

Figure 8.2 Not chemical bonding?

This same student described the interaction in 'part of a layer in a sodium chloride lattice' (Figure 8.2) as an 'attraction', but not a 'force', and neither as 'bonding' or a 'chemical bond'. A diagram showing 'iodine molecules in solid iodine' (Figure 8.3) was considered by this student to show 'van der Waal's [sic] forces'[3] – interactions which were 'force', but were not classed as 'attraction'. Another classmate thought that the interactions in the sodium chloride were 'bonding' but not 'chemical bonds'.

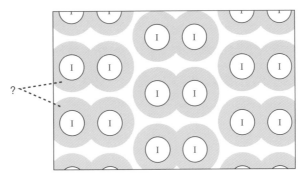

Figure 8.3 A force, but no attraction?

It is difficult to see any obvious pattern in these types of responses, except that where teachers may use terms such as 'attraction', 'force', 'bonding' and 'bond' without much thought (and often interchangeably), students are often drawing subtle and idiosyncratic distinctions.

Multiple models for chemical bonds

Chemical bonding is one of those topics where we use a variety of different models to understand different aspects of the phenomenon (see the discussion of models in Chapter 6).

For example, it is possible to teach about chemical bonds in terms of the electrical attractions between different species without mentioning orbitals – a topic that is often found difficult (see Chapter 7). Yet in post-16 level courses the idea of molecular orbitals, formed by the overlap and 'linear combination' of atomic orbitals is often introduced. A student who has learnt to conceptualise the covalent bond as a pair of electrons between, and attracted to, two nuclei, may find this new image of orbital overlap as something completely unrelated. Of course both of these 'pictures' are partial models of the same bond, but this will not be obvious to many students.

Again this is an area where helping students to appreciate that our descriptions and diagrams are just models will make learning easier.

Students' multiple conceptual frameworks for bonding

Just as teachers will use multiple models of bonding to help learners appreciate the abstract ideas involved, so students may develop manifold conceptions of chemical bonds. At some point, successful post-16 students are able to move beyond notions of bonds as shared electrons, to see bonds as electrical interactions. Those students who go on to study the subject at University level will be expected to master models of chemical reactions and bonding in terms of orbital interactions.

RS•C

Some of the alternative conceptions described in this chapter are very common among students. However, research suggests that the alternative ideas sometimes co-exist alongside developing more sophisticated understandings.[4] For example, it is often found that post-16 students are in transition between two models of the ionic bond (see Table 8.2). These students have two conceptual frameworks for making sense of the ionic bond – and may agree with (sometimes contradictory) statements that fit either framework.

The full shells explanatory principle may be a key part of a wider conceptual framework (the octet framework – see Chapter 1), where the different key ideas are mutually reinforcing. Unfortunately conceptual change can be very slow, and shifts in the preferred bonding explanations many take many months.[5] This is one area of the curriculum where students' alternative conceptions can be very tenacious indeed.

Student definitions of bonds: the bonding dichotomy

Research suggests that students at the end of secondary education commonly know about two separate categories of chemical bonding, and assume that chemical bonds must be like one of these two prototypes, (Table 8.1).[6]

Covalent	Ionic
Electrons are shared between non-metal atoms	Electrons are transferred from metal to non-metal atoms

Table 8.1 A dichotomy of bond types

Students seem to acquire this dichotomous classification of bonds readily, and when they do it means that they do not see bonding as primarily an electrical phenomena. Once this scheme has become established the student finds it difficult to appreciate bonding that is intermediate (polar bonds) or falls outside (eg hydrogen bonding) this narrow definition of bonding. Now clearly covalent and ionic bonds are very significant bond types, as many important substances can be understood to have – to a first approximation – either ionic or covalent bonding. However, the effect of pupils in school learning about bonding as a dichotomy of these two types, is to act as an impediment to later learning.

The covalent bond as electron sharing

Figure 8.4 A molecule with a covalent bond

From an electrical perspective the bond in a fluorine molecule (eg Figure 8.4) comprises a pair of electrons situated between the positive cores of the atoms. This arrangement does not make sense from a purely electrical standpoint – as negative electrons would not be expected to be paired as they will repel each other.[7] However few students question the 'pairing' of electrons, as they do not tend to see a covalent bond in electrical terms – rather they conceptualise the bond as a pair of electrons shared between two atoms to allow the atoms to have full electron shells (or octets).

Often, for the student, the bond is the sharing – and this is not necessarily meant figuratively (see Chapter 10). Students will report that atoms want to, and indeed need to, obtain full shells, and will therefore share to try to achieve this.[8] The teacher may talk of a shared pair of electrons as a shorthand for the electrical interaction – but to many students 'sharing' electrons is a technical and not a metaphorical description of the bond. It is therefore not surprising that some students

RS•C

completing the **Interactions** probe will classify covalent bonds as not being 'attractions' or 'forces'. Sharing is a 'social' process not a physical one!

The polar bond

Figure 8.5 A molecule with polar bonds

If a student does conceptualise bonding in electrical terms then a polar bond can be seen as something in-between a covalent bond and an ionic bond. The electron pair is pulled closer to one atomic core (or the electron density is greater nearer that core).

Where students think about bonding in terms of the dichotomy, however, they will tend to describe a polar bond as a modified covalent bond, rather than something intermediate between covalent and ionic. However, unless the bond polarity is drawn to their attention, it is quite likely they will ignore it completely. A diagram such as Figure 8.5 is likely to be just labelled as covalent, because the way it is drawn (see Chapter 7) fits a description of 'electron sharing' better than 'electron transfer'.

This tendency to ignore bond polarity leads to other errors. For example, as students tend to classify hydrogen fluoride as covalent, rather than polar, they often describe the solvated species to be hydrogen fluoride molecules when it dissolves in water. Some students also tend to assume that bond fission will always be homolytic (with each atom 'getting its own electrons back'). Appreciating heterolytic bond fission is easier if a bond is conceptualised as polar, in terms of electrical interactions, rather than as covalent with electrons shared. Terms like 'sharing' bring associations from normal social interactions (of fairness and collaboration in this case), whereas heterolytic bond fission would be in breach of such a social contract.

One of the classroom resources included in the companion volume is a probe called **Spot the bonding** which presents learners with a range of diagrams of chemical systems and asks them to identify the type(s) of bonding present.

When this probe was piloted for the project by a cohort of 16–17 year olds, it was found that there was virtually no explicit recognition of the polar nature of any of the bonds represented (even when the presence of hydrogen bonding or dipole-dipole interactions were reported).

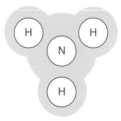

Figure 8.6 Representation of an ammonia molecule from Spot the bonding probe

For the ammonia molecule (Figure 8.6), 37 of the 39 students in the group identified covalent bonding – none suggested polar bonds (although there was one suggestion that the species had covalent bonding and 'permanent dipole-dipole forces').

RS•C

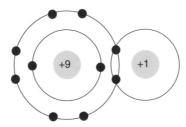

Figure 8.7 Representation of a hydrogen fluoride molecule from the Spot the bonding probe

In the case of the hydrogen fluoride molecule there were still no references to polar bonding, although 30 students suggested the bonding was covalent, and 4 suggested ionic. In the similar case of hydrogen chloride the split was 28 suggesting covalent and 7 suggesting ionic bonding. (Although chlorine is less electronegative than fluorine the type of representation – see Figure 8.8 – may have implied 'covalent' less strongly than in the case of hydrogen fluoride.) Even those students who suggested that there would be hydrogen bonding present (6/39), or dipole-dipole interactions (2/39), did not describe the bonding in the molecules as polar.

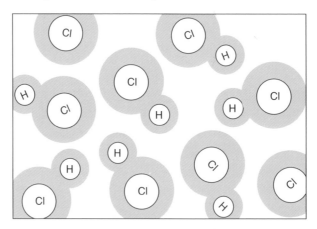

Figure 8.8 A representation of liquid hydrogen chloride from the Spot the bonding probe

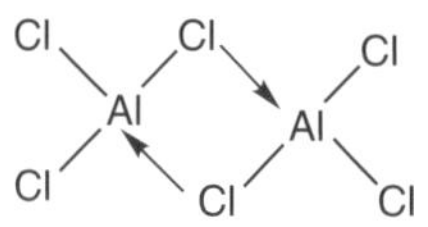

Figure 8.9 A representation of an aluminium chloride dimer from the Spot the bonding probe

Even in a case that students would find difficult to categorise according to the 'bonding dichotomy' they did not suggest the bonding might be polar. So in the case of aluminium chloride the bonds were drawn with lines (see Figure 8.9), in the way covalent bonds are often drawn, but the elements were a metal and a non-metal. 18 of the students suggested the bonding was covalent, and 11 suggested it was ionic.

Students are often taught that a dative bond is just like any other covalent bond, except that both electrons come from the same atom. This is meant to emphasise the important point that the origins of the electrons are irrelevant to their ability to bond together two atomic cores. Yet it can also obscure the fact that dative bonds are normally quite polar, with a very uneven 'sharing' of the electrons.

Bond polarity makes sense in terms of differences of electronegativity, for example in the hydrogen fluoride molecule – where the fluorine core charge is larger than the hydrogen core charge. If the

RS•C

bond is conceptualised in electrical terms (the electron pair is attracted to and by both nuclei), then understanding bond polarity may be seen as developing or refining existing knowledge. Yet this is not so for learners who see the bond as a shared electron pair per se. Students tended to see bond polarity as an additional secondary characteristic of covalent bonds (rather than as a continuum running between covalent and ionic).

The ionic bond as electron transfer

Students very commonly have alternative conceptions of ionic bonding. Where students often come to see the covalent bond in terms of an inadequate image (electron sharing), they often define ionic bonding in terms of a completely irrelevant notion: electron transfer.[9]

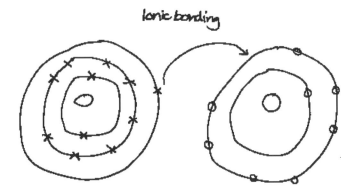

Figure 8.10 A student's diagram which has little to do with ionic bonding

Figure 8.10 is a student's representation of ionic bonding. This type of diagram is common, but – and I do not think this can not be emphasised too strongly – it has little to do with ionic bonding. Diagrams like Figure 8.10 certainly do not show ionic bonding. Rather they represent a way in which ions might be formed from isolated atoms.

Such diagrams of 'ion formation' have very little to do with the way ionic bonds are likely to be formed in school science. If a pupil could start with isolated atoms of sodium and chlorine then such diagrams might represent how these species would interact: but sodium is usually a metal, and chlorine is usually molecular. So if sodium chloride was prepared by binary synthesis the reactants would not be in the form of discrete atoms.

In any case students are much more likely to form sodium chloride by a neutralisation process, followed by evaporation of water. In such a reaction the ions are already present in the solution, and their formation does not need to be 'explained' to discuss the ionic bonding in the product. (See the discussion of precipitation reactions in Chapter 9.)

Unfortunately for many students ionic bonding is defined as 'electron transfer' by the time they compete formal schooling, even though this is completely wrong. The strength of this conviction must reflect the way the topic is presented (by textbooks if not by teachers – see Chapter 10). However, students seem to find it an easy image to learn as it fits with their wider understanding of the subject:

■ that everything is 'made from' atoms;

■ that bonds form to let atoms obtain full shells; and

■ that when atoms bond, something like a molecule is formed.

If covalent bonding is taught before ionic bonding, then it is not surprising that diagrams of 'ion-pairs' are commonly seen to be like molecules, or even to be molecules. Indeed at the end of formal schooling students are as likely to hold an alternative 'molecular' framework for understanding ionic

RS•C

bonding as to hold the scientific view. (More precisely, many students have a confused mixture of these ideas). This alternative framework is compared with the scientific ('electrostatic') view in Table 8.2.

	Molecular framework	Electrostatic framework
Status	Alternative framework	Curricular science
Role of molecules	Ion-pairs are implied to act as molecules of an ionic substance.	Ionic structures do not contain ion pairs – there are no discrete ion-pairs in the lattice.
Focus	The electron transfer event through which ions may be formed.	The force between adjacent oppositely charged ions in the lattice.
Valency conjecture	Atomic electronic configuration determines the number of ionic bonds formed, (eg a sodium atom can only donate one electron, so it can only form an ionic bond to one chlorine atom).	The number of bonds formed depends on the co-ordination number, not the valency or ionic charge (eg the co-ordination is 6:6 in NaCl).
History conjecture	Bonds are only formed between atoms that donate/ accept electrons, (eg in sodium chloride a chloride ion is bonded to the specific sodium ion that donated an electron to that particular anion, and vice versa).	Electrostatic forces depend on charge magnitudes and separations, not prior configurations of the system (eg in sodium chloride a chloride ion is bonded to six neighbouring sodium ions).
'Just forces' conjecture	Ions interact with the counter ions around them, but for those not ionically bonded these interactions are just forces, (eg in sodium chloride, a chloride ion is bonded to one sodium ion, and attracted to a further five sodium ions, but just by forces – not bonds).	A chemical bond is just the result of electrostatic forces – ionic bonds are nothing more than this (eg the forces between a chloride ion and each of the neighbouring sodium ions are equal).

Table 8.2 An alternative framework for ionic bonding[10]

One of the classroom resources included in the companion volume, **Ionic bonding**, allows teachers to test how strongly their students accept these two different frameworks for understanding ionic bonding. The probe asks student to judge the truth of statements relating to a diagram meant to show part of a layer of ions in a salt crystal (Figure 8.11). When this probe was administered to over 300 students in a range of schools and colleges it was found that 'alternative' statements were commonly judged true by students at the end of secondary schooling (who had studied the topic of ionic bonding), as well as by post-16 students before and after revisiting the topic during their chemistry classes.

RS•C

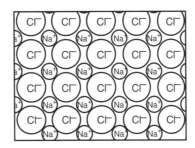

Figure 8.11 A focal diagram representing sodium chloride NaCl

For example the statement 'the reason a bond is formed between chloride ions and sodium ions is because an electron has been transferred between them' was considered correct by three quarters of respondents (69% of the sample of post-16 students who had revisited the topic, and 79% of each the other two groups). The statement that 'a chloride ion is only bonded to the sodium ion it accepted an electron from' was considered to be correct by almost half of the students (56% of the group at the end of schooling; 49% of the post-16 students relying on school knowledge; and 33% of the post-16 students having studied the topic at college level). Half the sample agreed that 'in the diagram a sodium ion is attracted to one chloride ion by a bond and is attracted to other chloride ions just by forces ' (52%, 53%, 45% of the different groups respectively.) A statement that 'there are no molecules shown in the diagram' was only judged as true by a third of the sample (27%, 33% and 37% respectively).

Is the metallic bond ionic, covalent, or just damp at the edges?

Metallic bonding is not part of the 'bonding dichotomy' (see Table 8.1) commonly used by students to categorise bonds, and is not usually studied in any depth before post-16 level courses. When students setting out on post-16 courses were asked about metals there was found to be four likely responses:[11]

■ there is no bonding in metals;

■ there is some form of bonding in metals, but not 'proper' bonding;

■ metals have covalent and/or ionic bonding; and

■ metals have metallic bonding, which is a sea of electrons.

In other words, many students tended to try and find a fit between the obvious structural integrity of metals, and their simple dichotomy of what bonding must be. For many of these students chemical bonding was understood in terms of striving to obtain a full outer shell by sharing (covalent bonding) or transferring (ionic bonding) electrons. As with ionic bonding, some students assume metals are molecular, and may try to think in terms of valency to make sense of diagrams they have seen in books (see Figure 8.12, which represents one college student's hybrid model based around the number of covalent bonds iron was considered to form).

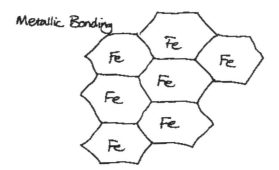

Figure 8.12 Metal molecules in iron (student diagram)

RS•C

Students who are unable to make sense of metals in terms of either pattern may be at a loss to explain how metals can have bonds: leading to comments such as 'it's not ionic, and it's not covalent either, it's like, it's hard to explain this'. These students may conclude that there is no bonding in pure metals, or that there is a 'lesser' form of bonding – just a force, and not a real chemical bond.

'Ionic and covalent bonds are formed ... where atoms lose or gain electrons, or share them, whereas metallic bonding is not the sharing or loss or gain of electrons. It is just a loose association with metal ions, and electrons they have lost'[12].

As discussed above, students often do not see bonds as being forces, and so may logically consider a structure to be held together just by forces, rather than by bonding.

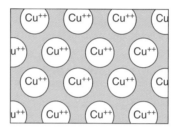

Figure 8.13 A representation of copper in the Interactions probe

The **Interactions** probe included in the companion volume includes an item about copper. One student who labelled the interaction present as 'metallic bonding' thought that this counted as both an 'attraction' and a 'force'; but was unsure if it could be considered as 'bonding', and did not class it as a 'chemical bond'.

Other students find ways to understand the metallic bond as a variation on the ionic or covalent case, with electrons being conceptualised as being shared, or being moved around so that the atoms take turns in having full shells (either by gaining enough, or losing enough electrons). A more appropriate model sees the electrons as having been transferred to the lattice so that the atoms could gain a full shell.

Although some students will describe the bonding in metals in terms of the 'sea' of electrons, they have often learnt the term by rote, with little understanding of this model (see Figure 8.14). Some student diagrams show the 'sea' as a vast excess of electrons (Figure 8.15), and one student who grasped that metallic bonding was 'the attraction between the +ve charge of the metal ions and the –ve charge of the electrons' went on to add that 'it also has a sea of electrons which flow around the structure.'

Like the case with the term 'sharing', the sea metaphor brings its own associations, with students referring to the metal cation 'being like an island surrounded by electrons' and 'floating in a sea of delocalised electrons'. The use of metaphors in teaching science concepts is considered in Chapter 10.

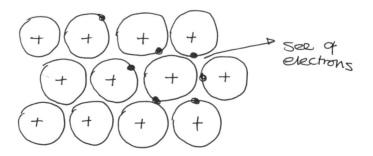

Figure 8.14 Can you 'see' the electrons? (student diagram)

RS•C

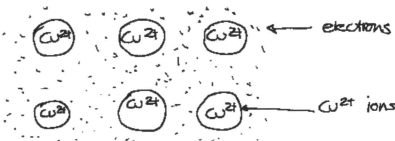

Figure 8.15 An electronegative metal? (student diagram)

Included in the companion volume is a resource, **Iron – a metal**, which allows teachers to diagnose some of the common alternative conceptions that students may have about metals and metallic bonding. Reference was made in Chapter 6 to how students in one group of 14–15 year olds confused the properties of the metal with properties of metal atoms. Other items in the probe allow teachers to diagnose students' ideas about the bonding in a metal. Half (9/18) of the same group of 14–15 year olds thought that 'the atoms [*sic*] in a metal such as iron are held together by ionic bonds'.

Intermolecular interactions as bonds, or just forces?

If many students have difficulty thinking of metallic bonding as proper chemical bonding, then the status given to intermolecular bonding tends to be even less firm. It is common for these interactions to be considered as 'just forces' rather than as 'proper' bonding. Although this may sometimes be a distinction based upon the strength of the interaction (*eg* van der Waals forces are too weak to be considered 'proper' bonds), this is not a full explanation when the interactions in an ionic lattice are often divided into 'ionic bonds' (resulting from electron transfer) and 'just forces' (due to electrical attraction between ions). For many students the criterion for a bond is not related to the strength of the interaction, but to whether the bond results in atoms obtaining full shells.

The full shells explanatory principle for explaining chemical bonding, that is that bonds are formed 'to try to achieve a stable structure *ie* 8 electrons in the outer shell of the atom' is a very common alternative conception. Although it may be expressed in various ways ('to give them all full outer shells', 'to obtain the configuration of a noble gas, and be stable', 'to form a completed outer shell'), the underlying idea is virtually ubiquitous. If a student understands bonding purely in these terms then forms of intermolecular bonding such as van der Waals forces and hydrogen bonding will not qualify, and so will be classed as 'just forces'.

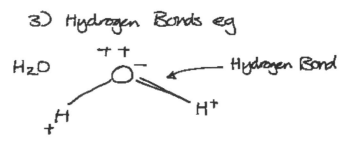

Figure 8.16 A bond to hydrogen (student diagram)

Hydrogen bonds may be misunderstood by students as simply being 'bonds to hydrogen' (see Figure 8.16) and just thought of as covalent bonds. This sometimes occurs when students are introduced to hydrogen bonding, without further elucidation, in biology before they have met the concept in chemistry (see Chapter 10).[13]

RS•C

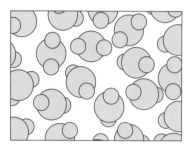

Figure 8.17 A representation of molecules in liquid water, from the Interactions probe

When students realise that this is not what chemists mean by hydrogen bonds they may often be demoted from being real bonds. Although hydrogen bonds are electrical interactions (and involve orbital interactions) they do not allow atoms to obtain full shells, which is the most common criterion used by students to characterise a chemical bond. For example, a student responding to the **Interactions** probe included in the companion volume labelled the interactions between molecules in water (Figure 8.17) as 'hydrogen bonding'. The student described how:

'The highly electronegative oxygen in H_2O attracts each hydrogen's electron so strongly that we basically have an exposed H nucleus in the H_2O molecules giving the H a permanent $\partial+$ and the Os a permanent $\partial-$ charge. The neighbouring molecule $H^{\partial+}$ are attracted to the $O^{\partial-}$, thus holding the molecules together in the liquid.'

This student certainly appreciated the electrical nature of the interaction, but did not think that this amounted to a chemical bond.

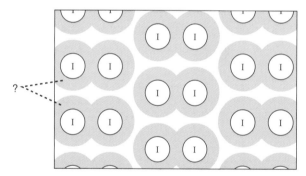

Figure 8.18 A representation of molecules in solid iodine, from the Interactions probe

In a similar way a student who recognised 'van der Waals forces' in Figure 8.18, where 'molecules [are] attracted by induced dipoles which are instantaneous and attract molecules together' thought this did not count as 'bonding' nor as a 'chemical bond'. The same student described the 'electrostatic attraction' that gave rise to solvation (Figure 8.19) where 'oxygen in water attracts Ag^+ and H in water attracts NO_3^-, but again this was not categorised as 'bonding' or as being a 'chemical bond'.

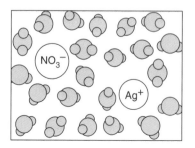

Figure 8.19 A representation of solvation from the Interactions probe

RS•C

In these cases the students seemed to have a fair understanding of the interactions, even if they did not consider them as bonds. Indeed it may well be the label 'bond' which stands in the place of understanding for some students – the same individual who was able to describe van der Waals forces in iodine in terms of induced dipoles and solvation in terms of attractions, described the bond in a hydrogen molecule as a 'shared pair of electrons between the two hydrogen atoms in a covalent bond', without any reference to what made this hold the molecule together.

However, the students who completed the **Spot the bonding** probe often failed to spot intermolecular bonding in the diagrams represented. Only 14/39 students suggested there would be any type of intermolecular bonding in solid iodine; 15/39 in liquid oxygen; 15/39 in liquid hydrogen chloride and 16/39 in sodium nitrate solution.

Developing understanding of chemical bonding

So to summarise, many students define bonding as a way of getting full shells (by electron sharing or electron transfer). Students who think this way may tend to ignore other types of bonding which do not fit this scheme (polar bonds, metallic bonds, intermolecular bonds), and may be slow to move beyond inadequate definitions of covalent and ionic bonding.

However, this does not mean that students can not move beyond their limited models of the bond. The octet framework (see Chapter 1) is a substantive learning block (see Chapter 4) to many students, but is not necessarily insurmountable. Students will use multiple conceptual frameworks, as discussed earlier in this chapter, and successful post-16 students do slowly learn to adopt models of chemical bonds in terms of electrical forces and even orbital interactions, and can be gradually weaned away from relying on the full shells principle.

Some of this may happen without any deliberate intervention by teachers. For example, students who had studied Raoult's law and had been taught that solutions deviated from ideal behaviour when the mixture had different types of bonding to the pure liquids were found to refer to intermolecular bonding and solute-solvent interactions as being bonds in that context.

As students often have a poor appreciation of the nature of models they may need to be encouraged to accept that they can use alternative types of explanations. One way of doing this is to encourage students to see their chemical ideas as components of a mental toolbox from which they need to select the best tool to tack a particular 'job' (explanation, examination question *etc*).[14] Students may feel more comfortable with acquiring and trying out manifold conceptions if they see explanations as being like stories – where they must find the best fit between the phenomena and the explanation.

Figure 8.20 represent this figuratively. This diagram uses a 'profile' to show the extent to which different types of bonding explanations were given when a student on a two year post-16 chemistry course was asked about aspects of chemical bonding.[16,17,18] In this diagram the 'target' is an understanding of chemical bonding based around electrical and some orbital ideas, but the student commences the course habitually using the idea of 'full shells' as the main way of discussing bonds. During the course this student developed a new principle for bonding (that 'bonds form to minimise energy') as well as the intended ideas. By the end of the course the student had moved some way towards the way of thinking that the teacher was trying to encourage, but still commonly used 'full shell' explanations, which were so well established.

RS•C

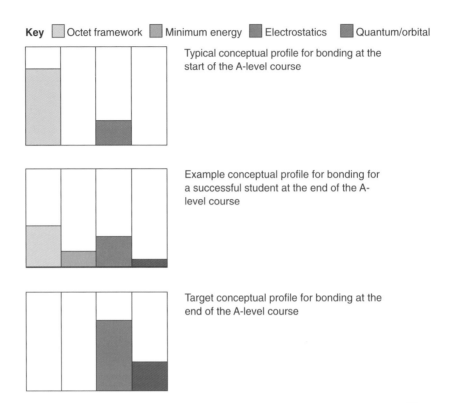

Key ☐ Octet framework ■ Minimum energy ■ Electrostatics ■ Quantum/orbital

Typical conceptual profile for bonding at the start of the A-level course

Example conceptual profile for bonding for a successful student at the end of the A-level course

Target conceptual profile for bonding at the end of the A-level course

Figure 8.20 Progression in understanding bonding during post-16 chemistry[15]

It is worth noting that in Figure 8.20 the 'minimum energy' principle is shown as a discrete type of explanation. From a scientific point of view the idea that bonds form as chemical systems evolve to a lower energy is closely related to the idea that electrical forces act to give chemical structures at equilibrium. The 'lowest energy' point is the equilibrium configuration where the forces are balanced.

However, for the student who is profiled in the diagram, the 'minimum energy principle' was a completely separate idea. Once again a student failed to appreciate what the teacher took for granted, and so – for this learner – explanations in terms of minimising energy, and those in terms of electrical forces, were stored and accessed as separate stories for explaining bonds. This fragmentation of knowledge (see Chapter 4) was an impediment to the student developing a deeper understanding, but given more time and more experience of using his different stories it is quite possible this student may have come to integrate these different accounts.

Teachers who wish to encourage their students to develop their ideas about chemical bonding should:

■ emphasise the nature of bonds as electrical interactions;

■ avoid using anthropomorphic language, but rather explain bonding in terms of forces; and

■ avoid talking about electron transfer (ion formation) when considering ionic bonding.

Notes and references for Chapter 8

1. K. S. Taber, An alternative conceptual framework from chemistry education, *International Journal of Science Education*, 1998, **20** (5), 597–608.

2. Or, to be precise, it is an equilibrium of forces, which holds the chemical system in balance; and which means that work has to be done to move the system from that equilibrium.

3. Students' spelling of van der Waals can be somewhat variable. In the responses one group made to the **Spot the bonding** probe the variations presented included: Van der waals forces, Vanderwaals, Wan der Waals' forces, Wam der Vaals, van der walls and van der balls forces.

RS•C

4. K. S. Taber, Multiple frameworks?: Evidence of manifold conceptions in individual cognitive structure, *International Journal of Science Education*, 2000, **22** (4), 399–417.

5. K. S. Taber, Shifting sands: a case study of conceptual development as competition between alternative conceptions, *International Journal of Science Education*, 2001, **23** (7), 731–753.

6. K. S. Taber, An alternative conceptual framework from chemistry education, *International Journal of Science Education*, 1998, **20** (5), 597–608.

7. The concept of 'quantum mechanical spin' is not normally introduced in secondary level chemistry.

8. K. S. Taber & M. Watts, The secret life of the chemical bond: students' anthropomorphic and animistic references to bonding, *The International Journal of Science Education*, 1996, **18** (5), 557–568.

9. K. S. Taber, Misunderstanding the ionic bond, *Education in Chemistry*, 1994, **31** (4), 100–103.

10. K. S. Taber, Student understanding of ionic bonding: molecular versus electrostatic thinking?, *School Science Review*, 1997, **78** (285), 85–95.

11. K. S. Taber, Building the structural concepts of chemistry: some considerations from educational research, *Chemistry Education: Research and Practice in Europe*, 2001, **2** (2), 123–158, available at **http://www.uoi.gr/cerp/** and **http://www.rsc.org/Education/CERP/index.asp** (accessed September 2005).

12. K. S. Taber, Understanding Chemical Bonding, unpublished PhD thesis, University of Surrey, 1997, 354.

13. K. S. Taber, Understanding Chemical Bonding, unpublished PhD thesis, University of Surrey, 1997, 838–839.

14. K. S. Taber, An analogy for discussing progression in learning chemistry, *School Science Review*, 1995, **76** (276), 91–95.

15. This figure is taken from K. S. Taber, An explanatory model for conceptual development during A-level chemistry, 1999, available via Education-line, at **http://www.leeds.ac.uk/educol/** (accessed September 2005).

16. This is an approach based on the ideas of the French philosopher of science, Gaston Bachelard (see note 17) and proposed as a useful technique to follow conceptual development (see note 18).

17. G. Bachelard, Gaston *The Philosophy of No: A Philosophy of the Scientific Mind*, New York: Orion Press, 1968.

18. E. F. Mortimer, Conceptual change or conceptual profile change?, *Science and Education*, 1995, **4**, 267–285.

RS•C

RS•C

9. Chemical reactions

The study of reactions is at the core of chemistry as a subject. This chapter considers areas where students are known to have difficulties in learning about reactions in chemistry. It also introduces some of the classroom exercises included in the companion volume to help teachers elicit and challenge their students alternative conceptions.

Describing chemical reactions

The main defining characteristic of a chemical reaction[1] is that it is a process where one or more new substances are formed, *ie*:

$$reactants \rightarrow products$$

Here the reactants and products are different chemical substances.[2,3] The basic way of describing chemical reactions, then, is to write 'chemical equations' showing the reactants and products in a particular chemical change. Chemical equations have been described as 'an essential part of the common language of scientists'.[4]

Chemical 'equations' are commonly written as word equations and formulae equations. There is a sense in which formulae equations are easier, as they provide a ready means of checking that no transmutation has been implied (by seeing that the same elements are represented on both sides of the equation). However, formulae equations are more abstract, and so word equations are often introduced first.[5]

Word equations relate to the macroscopic level of laboratory phenomena that may be directly experienced by students, whereas formulae equations are directly related to the molecular level (see Chapter 6). The full implications of formulae equations can only be appreciated by a student who has been introduced to the molecular and/or ionic nature of the reacting substances.[6]

Although word equations use the (often) less abstract names of substances, rather than formulae, they can make it more difficult for students to check that the same elements are represented before and after a reaction. Students need to know, for example, that the ending '-ate' implies the presence of oxygen, and which elements are present in common substances such as water or ammonia.

Student difficulties with word equations

Word equations are commonly introduced and used in lower secondary science, but national testing of students at this age in England and Wales shows that many students find it difficult to write or complete word equations for chemical reactions.[7]

This should not be surprising if it is remembered that although chemical names seem familiar to teachers (with many years of acquaintance using them) they may seem somewhat arbitrary to students. One science educator has told me how in school he spent two years perplexed at why chlorides should be produced in reactions of 'hydraulic' acid! The more systematic names, such as tetraoxosulfate(VI) for the SO_4^{2-} ion, may be especially difficult for students.[8] If the names do not seem to fit an accessible pattern, then students may well be concerned about the very large number of substances they could hear and read about.

Clearly the expert (the teacher) has another advantage in having a well established framework for seeing specific reactions as particular examples of common reaction types. For the teacher, this classification system acts as a set of familiar 'mental pigeon-holes' into which reactions may be slotted. Each of these 'slots' is a ready made set of mental connections for the specific reaction being discussed (see Figures 9.1 & 9.2).

RS•C

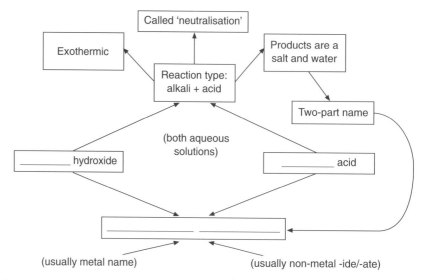

Figure 9.1 A 'mental slot' ready for thinking about a type of reaction

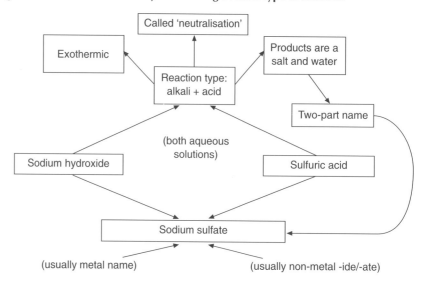

Figure 9.2 A specific example is fitted into the 'mental slot'

Students, lacking this framework for organising the information, see the equations at a much finer resolution – in the terms of Chapter 5 – so that a chemical equation that the teacher perceives as an integral unit seems to the students to have many components. For the expert (the teacher) the equation is easier to hold in mind once it is recognised as fitting a familiar 'slot'. For the novice (the student), lacking the teacher's background, classifying the reaction is itself an additional cognitive demand, which does not automatically bring the benefits that accrue to the expert.

One of the resources included in the companion volume, **Word equations**, is an exercise on completing word equations. This exercise is not designed around teaching students a set of rules, but rather with giving them an opportunity to see that, although there are many possible chemical reactions, those met in school science may often be fitted into a limited number of common types. (The types used in this resource are binary synthesis; displacement of a metal from a salt; neutralisation; acid with carbonate; and acid with metal.)

This approach is designed to help students start constructing the reaction-type mental 'slots' into which they can 'drop' reactions (in the way discussed above).

The teaching exercise is accompanied by two diagnostic assessment probes, which ask students to complete word equations. Whereas the teaching exercise presents the examples within a framework with plenty of cues (to help students work out the answers), the two assessment tasks simply present five incomplete word equations for students to complete. (The teaching exercise may be seen as providing 'scaffolding', in the sense discussed in Chapter 5, a structure within which the students are enabled to complete a task which might otherwise be beyond them.)

When the materials were piloted the two assessment tasks were used as a pre-test and post-test to enable teachers to judge if students had learnt anything by undertaking the teaching exercise. This would be a sensible way of using the materials if they are adopted with a class that should have already covered the material. (Indeed, if the students demonstrate competence on the pre-test, then the exercise and post-test will not be needed).

However, if the materials are used as part of the normal introduction and teaching about word equations then it would be more appropriate to use both of the diagnostic assessment probes as post-tests after the teaching exercise. This could either be as an immediate and a delayed post-test (one used straight after teaching, another some time later when revising the topic), or the first could be used as a 'filter', with only those having difficulties asked to undertake some more work and the second probe.

When these materials were piloted it was found that many students found difficulty in completing simple word equations (something which had been highlighted previously in national testing of 14 year olds in England and Wales, and so had been anticipated). It is worth considering some of the responses to the pre-test, from students in a group of 14–15 year olds.

Perhaps the simplest question was the item based on a binary synthesis. Most students realised that a reaction between calcium and chlorine would produce calcium chloride, although calcium chlorate was also suggested. Similarly, many students were successful on the item based on the displacement of a metal from its salt by a more reactive metal. It was generally recognised that it would need to be copper nitrate (solution) which reacted with zinc to give zinc nitrate (solution) and copper. Even here there was some alternative answers. One student suggested the answer was copper sulfate because 'zinc will displace copper sulfate'. This response seemed to be based on trying to recall the reaction, rather that work out the logic of the equation.

Another student in the class who did try to use the information given produced an incomplete answer of nitrate as 'it says zinc nitrate'. This student did not bring to mind that the question 'also said' copper. However, in general, these two items seemed to be tackled well by the group.

The other three items seemed to place heavier demands on the students. The neutralisation item required students to identify the salt (ie potassium nitrate) which would be produced from a particular acid (nitric acid) and a particular alkali (potassium hydroxide). This question produced a range of alternative suggestions from this one teaching group: nitric hydroxide; hydrochloric acid; potassium nitric acid; hydrogen; potassium nitrate acid; nitric oxide; potassium acid; nitric acid. Whilst some of these responses were guesses (as admitted by the students), others were genuine attempts to apply logic based on the information given, and the type of things that happen in chemical reactions,

'nitric acid + potassium hydroxide → potassium nitrate acid + water

[because] when water is taken out of the equation you are left with potassium nitrate acid.'

'nitric acid + potassium hydroxide → potassium acid + water

[because] potassium will displace the nitric acid'

The acid which is reacted with zinc carbonate to produce zinc sulfate, water and carbon dioxide was suggested to be sulfate acid; hydrochloric acid; hydroxide acid; and sulpher [sic] hydroxide acid. Again there was some evidence of a logical approach,

'hydroxide acid + zinc carbonate → zinc sulfate + water + carbon dioxide

RS•C

[because] when looking at the [equation] I can tell what is missing from either side as it is meant to be equal.'

This student was partially right: yes, it was meant to be 'equal', but, no, he couldn't tell what was missing.

The item based on a metal reacting with an acid also gave a range of responses. When magnesium reacts with hydrochloric acid to produce hydrogen, then the other product expected by various students was water; magnesium hydroxide; magnesium oxide; magnesium – because 'it is magnesium and hydrochloric acid'; and magnesium acid 'because I believe the acid has been given to the magnesiume [sic]'.

What this brief examination of the responses of a single group suggests is that while some students are totally at a loss with word equations, others make mistakes despite having a valid strategy. Some of those students who appreciate the basic idea of conservation (ie both sides are equal in some sense), and realise that they should be able to work out 'what is missing', still fail because they are not familiar enough with the naming of categories of chemicals (oxide, hydroxide, acid, -ate, -ide, etc) or with the general equations for reactions (the mental pigeon holes discussed above). The demands prove too great for students who can not effortlessly call upon these mental resources. The task is too complex when seen at the resolution available to the student.

Types of reactions

Only a limited number of types of reactions are commonly considered during school science. However, these categories are not mutually exclusive, which may confuse students.

One of the probes included in the companion volume, **Types of chemical reaction**, presents a range of reactions to students, and asks them to classify the reactions in terms of five specific categories and a 'none of the above' option. The examples chosen are intended to be those that students are likely to meet in their courses, and are described in both word and formulae equations. The students are also asked to explain their choices, so that they will think about their own reasons for making the classifications. Some students will find this quite difficult, but this 'metacognition' - thinking about thinking (see Chapter 3) - may be useful when asking students to revisit any question that have classified incorrectly. It may also help teachers to spot why students are making mistakes – a key aspect of diagnostic assessment (see Chapter 10).

The categories used are displacement, neutralisation, oxidation, reduction and thermal decomposition. The decision to give separate categories for oxidation and reduction, rather than a 'redox' category, was based upon the way these concepts are often introduced (in terms of addition and loss of oxygen and hydrogen). Clearly, when the idea of redox is taught, it becomes an appropriate teaching point that whenever a reaction is classified as an oxidation it must also be a reduction, and vice versa.

When this probe was piloted, it was found that students had considerable difficulty in classifying reactions. One aspect that clearly troubled some students was that the same reaction might be an example of several categories of reaction.

For example, the first question asks about the reaction between nitrogen and hydrogen to produce ammonia. This is a redox process, and even students operating with an elementary concept of oxidation and reduction might be expected to consider nitrogen as being reduced (due to the addition of hydrogen). However, a number of students in a class of 15-16 year olds classed this reaction in the 'none of the above' category. The reasons that students gave were often that it fitted some other category that was not on the list. The reaction was described variously as:

■ a reversible reaction;

■ a simple reaction of two elements;

■ requiring heat;

RS•C

- a normal reaction; and, by contrast,

- an unusual reaction.

One student also put the electrolysis of sodium chloride in the 'none of the above' category because 'it is electrolysis of brine'. It would seem that when some students already had a label for the reaction and found that it was not listed, they did not examine the other categories closely to see if they also applied. This is a particular concern when many reactions may be classed as displacement reactions. For example the reaction between sodium hydroxide and nitric acid was considered to be a displacement (and in no other category) as 'the Na is more forming a new chemical with the nitrate [because] they are more reactive' or because 'they swap around'. Similarly the reaction between copper carbonate and sulfuric acid was also considered to be a displacement 'because the CO_2 has been displaced' and because 'the substances displaced each other'.

Other students thought that this reaction was a neutralisation 'because the sulfuric acid has been neutralised'. Although this is an acid-base reaction, the term neutralisation is usually reserved for acid-alkali reactions. Another student in the group classed the reaction between zinc (a 'base' metal) and hydrochloric acid as a neutralisation.

This was not the only unconventional interpretation of a term. The decomposition of copper carbonate into copper oxide and carbon dioxide was classed as a reduction 'because the copper carbonate has been reduced to form copper oxide and carbon dioxide'. The same individual also classed the electrolysis of sodium chloride in this way 'because the sodium chloride has been reduced to sodium and chlorine'. This would seem to be a sensible application of the everyday 'life-world' meaning of 'reduced' (see Chapter 2).

The combustion of methane was classified in two different ways.[9,10,11] Some students saw this as a displacement reaction as 'oxygen take[s the] place of hydrogen in carbon' (or 'the oxygen displaces the hydrogen'), or, as another student explained, 'they swap around'. Other students saw this reaction as a thermal decomposition. One student pointed out that 'the steam indicates [that] there's heat', and another explained that 'the heat is put into the methane and oxygen then it will form carbon dioxide and steam'. Another classmate made the valid observation that 'the methane has been broken down by heat', but – of course – chemists use the term thermal decomposition in a more restricted sense.

Why do reactions occur?

Explaining 'why' reactions occur is a much more difficult task than either completing equations or classifying reactions. Indeed, this topic is not usually dealt with in any meaningful sense until at post-16 level. In one sense this is understandable, as valid explanations require detailed consideration of bond enthalpies, or – at least – a sophisticated application of notions of electrode potentials and Gibbs free energy. In some cases a qualitative treatment of entropy can be used.

Yet research has suggested that many students believe that they do know why reactions occur by the time they complete their secondary science courses.[12] However, the reasons they commonly give are invalid, and may be in contradiction to ideas that they would be expected to learn if they continue their study of chemistry to post-16 level.

The most common alternative conception is to suggest that chemical reactions occur 'so that atoms can acquire a full outer electron shell' or 'an octet of outer electrons'. Students at the end of secondary schooling, or during post-16 courses, will commonly give a response along these lines, even if the information given in the question clearly shows this cannot be a valid explanation.

One of the resources included in the companion volume, **Hydrogen fluoride**, is a diagnostic probe which enables teachers to explore their students' understanding of why reactions occur. In this diagnostic probe students are told that 'Hydrogen reacts with fluorine to give hydrogen fluoride.' The students are given both the word and formulae equations to consider.

RS•C

The equation for this reaction is:

$H_2(g) + F_2(g) \rightarrow 2HF(g)$

The word equation is:

hydrogen + fluorine → hydrogen fluoride

Students are also presented with a diagram (see Figure 9.3).

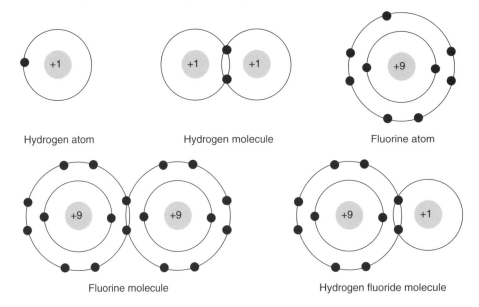

Figure 9.3 Figure used in the Hydrogen fluoride probe

The task students are set is to 'In your own words, explain why you think hydrogen reacts with fluorine'. The figure shows a molecule of each of the reactants, and of the products, and also of the hydrogen atom and the fluorine atom. At the molecular level, the process which needs to be explained is that given in Figure 9.4.

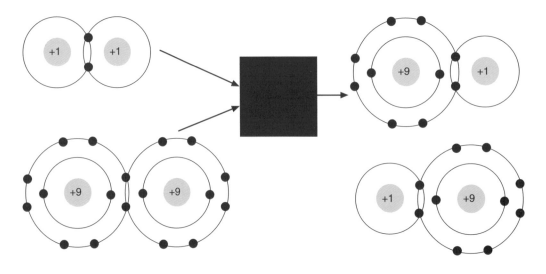

Figure 9.4 The reaction between hydrogen and fluorine

RS•C

Both the reactants and products are molecular, and so any explanation that students produce at the molecular level needs to relate to these species (as is implied by the equation showing the reactants as $H_2(g)$ and $F_2(g)$). In this case entropy cannot be used as an explanation.

The figure presented to students also includes the atoms, as these are known to often dominate student thinking (see Chapter 6). Of course, atoms could be involved in the mechanism (in the black box in Figure 9.4), as could a number of other species (see Figure 9.5). However, the reaction process can not be explained in terms of properties of isolated atoms, as the reactants are not present in this form. I have laboured this point, because in many groups it is found that this is how students explain the reaction.

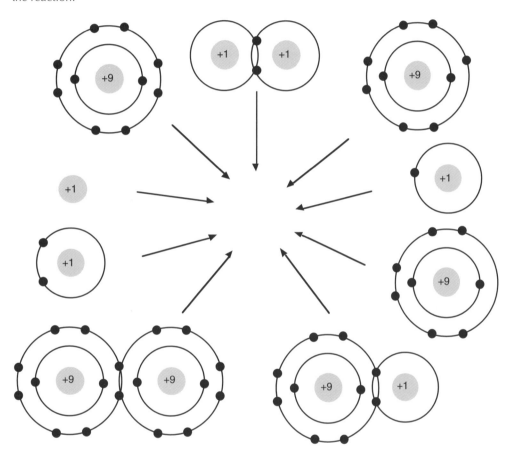

Figure 9.5 Some species that could be present (at least transiently) as hydrogen and fluorine react

When this probe was piloted, it was indeed found that many students explained the reaction in terms of the atoms of hydrogen and fluorine striving for octets of electrons. (See Chapter 8 for a consideration of how students apply the 'full shell explanatory principle' to bonding.)

For example the following response was from a student in a group of 14–15 year olds:

'All atoms have electrons in their outer shells. They want to get a full outer shell. Fluorine atoms need one electron to get a full outer shell, and hydrogen only has one, but needs one more. Therefore, they bond together, the hydrogen electron fills in the gap in the fluorine, and the hydrogen uses one of the fluorine electrons so they both have full shells.'

This was far from an isolated suggestion. The following response was from a student in a group of 15–16 year olds in another school,

'Hydrogen contains only one electron. The first shell of an atom always contains only two electrons at the most, while the other shells following that can contain up to eight electrons each. This means

RS•C

that in order to reach a stable state hydrogen needs only to gain an extra electron to have a full outer shell. Fluorine has seven electrons in its outer shell so needs to gain one more too. A hydrogen and a fluorine atom both share an electron each to form a single covalent bond.'

Both of these explanations are clearly from eloquent and thoughtful students. The are both 'logical' explanations, but based on the premise that the reactants are present in the form of atoms. To be fair to these students (and many others in these classes making similar suggestions), the issue of why reactions occur is seldom taught in any depth (if at all) in secondary school, and many student texts do imply that reactions occur to allow atoms to fill their shells (see Chapter 10).

Once students study chemistry in post-16 courses they will study concepts such as bond enthalpy, free energy changes, and reaction profiles. However, even students who have studied these topics may tend to explain the reaction in terms of the 'needs' of discrete atoms. Students in a group of 17–18 year olds gave the following explanations when responding to the probe:

'A fluorine atom has an incomplete outside shell of electrons of only 7, and a hydrogen atom also has an incomplete outside shell of 1....'

'...hydrogen reacts with fluorine as fluorine only has 7 electrons in its outer shell and needs hydrogen's single electron to give it 8 electrons in its outer shell and make it a stable molecule ...'

'Because both atoms need one extra electron in their outside shell to have a noble gas structure, so by sharing 2 electrons (one from each atom) in a covalent bond, hydrogen fluoride becomes a very stable molecule....'

'Hydrogen and fluorine atoms are both uncharged (but unstable) particles with unfilled outer shells of electrons....'

Similar responses to this probe, starting from the electronic configurations of isolated atoms, have been obtained from many students in a range of institutions. This would seem to be a very prevalent and strongly held alternative conception, making up a part of the common 'octet' alternative conceptual framework (see Chapter 1).

How do reactions occur at the molecular level?

Given that the 'full shells' explanatory principle can not be called upon to explain why reactions occur, there is a need to provide students with an alternative explanation.

At post-16 level a thermodynamic approach is often taught. In this the main considerations are the usually the bond enthalpies of the bonds broken and formed in a chemical process. In the example of the reaction between hydrogen and fluorine, bonds are broken in hydrogen (requiring 436 kJ mol^{-1}) and in fluorine (158 kJ mol^{-1}), and formed in hydrogen fluoride (562 kJ mol^{-1}, bearing in mind that 2 moles of HF are produced for each mole of hydrogen(/fluorine) reacting).[13] If there is an increase in disorder than entropy effects can also be considered.

An explanation may therefore be given in terms of the products being more stable, as there are stronger bonds in the product. If students require a mental image, then they can be asked to think about the electrical interactions between nuclei (or atomic cores) and valence electrons when the molecules interact. Those students taking post-16 level courses in physics may be able to think about the molecules quantacting (see Chapter 6) in terms of lowering the electrical potential of the system of charges. However, this is a rather abstract notion, and a deep understanding requires students not only to bring in ideas from physics (which many find difficult – see Chapter 7) but also to bear in mind that quantum theory places restrictions on the allowed configuration of electrons and nuclei.

Nevertheless, it seems sensible to encourage students to think about reactions in terms of the quantaction of the molecules, and the electrical interactions between different reacting species. This will certainly be useful when students are expected to appreciate and explain reaction mechanisms (see below).

RS•C

Practice in this area may be important, as evidence from research suggests that post-16 level students may often have difficulty visualising molecular level processes that their teachers may think are quite straight forward.

Consider, for example, the case of precipitation (double decomposition) reactions, such as that which forms the basis of the common test for chloride ions:

sodium chloride(aq) + silver nitrate(aq) → sodium nitrate(aq) + silver chloride(s)

The basis of this reaction seems simple. In the two solutions the ions present are all solvated (*ie* hydrated), but when the ions are mixed the silver cations and chloride anions bond together, and precipitate as silver chloride. An explanation of this may be given in terms of the energy involved (released) in solvating the various ions, compared with the energy involved (released) in forming the various possible crystal lattices ($AgNO_3(s)$, $NaCl(s)$, $AgCl(s)$, $NaNO_3(s)$). In terms of the 'molecular' level species present, the quantaction process concerns the strong forces of attraction between silver ions and chloride ion bringing them together to the exclusion of the solvating water molecules.

Although, no doubt, some transient clumps of other ion combinations also occur in the melee, the random motion of the water molecules is presumably vigorous enough to break up these groupings – which is why the other three compounds involved ($AgNO_3$, $NaCl$, $NaNO_3$), unlike silver chloride, are fairly soluble.

However, when post-16 level students have been asked about such reactions, they may describe a very different mental image of the quantaction process – suggesting that the basis of the reaction is silver atoms donating electrons to chlorine atoms to form the ionic bond.

In part this 'explanation' is just one manifestation of the tendency for students to think of reactions in terms of atoms (see the example of the reaction between hydrogen and chlorine, above, and the discussion of the 'atomic ontology' in Chapter 6), and to identify ionic bonding with electron transfer (see Chapter 8). However, this also demonstrates that many students do not have a very clear mental image of the particles present in a solution.

Students' ideas about the mechanism of ionic precipitation

One of the resources included in the companion volume, **Precipitation**, includes a diagnostic probe designed to elicit students' ideas about precipitation reactions. This is based around the example of precipitating silver chloride (see above). The reaction might be summarised as in Figure 9.6.

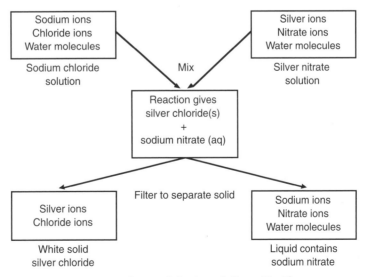

Figure 9.6 The precipitation of silver chloride

RS•C

The probe presents students with figures showing the particles in silver nitrate (solid), sodium chloride (solid), water (liquid) and silver chloride (solid). The particles shown in each diagram (specific ions or molecules) are also listed. The probe comprises of a structured question which asks about the particles present when solutions of each of sodium chloride and silver nitrate are formed (see Chapter 6); the particles in the liquid after the reaction (when the solid has been filtered off); and the bond in the precipitate.

When this probe was piloted in schools and colleges it was found that many 14–16 year old students have genuine problems making any real sense of the precipitation process. For example, in one group of 14–15 year olds, the explanations of what happens when the ionic bond forms included:

- 'The ions lose their individual charges and stick to each other.'

- 'The sodium and silver ions have switched places because the sodium ions are more reactive and therefore a displacement reaction will take place.'

- 'The silver will gain an electron and the chloride will lose an electron to become Ag and Cl instead of Ag^+ and Cl^-. They become stable.'

- 'The solid silver chloride is formed and hydrogen is given off along with silver chloride which is a solid.'

- 'The outer electron from the silver transfers into the outer shell of the chlorine. There would be an increase in the number of bonds.'

- 'The silver and chloride ions join together to form one solution.'

- 'They become neutral atoms as the charge on the ions cancel each other out and are no longer ions, they are atoms.'

- 'I think that it forms by silver giving some or one of its outer electrons to the chlorine + so they both end up with full outer shells. Then they bond together with intermolecular forces.'

- 'They both share electrons as it is mutually beneficial to the both of them.'

- 'It changes to covalent bonding'.

- 'The two atoms react together and there [sic] charges go because they both have a charge of 1 so then the molecules become neutral.'

- 'They separate and break away from each other.'

Some students in the same group did produce more acceptable answers:

'I think that as silver (Ag^+) will be attracted to the chloride ions (Cl^-) making the compound have very strong intermolecular [sic] bonds as it is very difficult to separate a positive and a negative charge.'

'Heat is given off as forming bonds is an exothermic process and the ions are attracted towards each other by electrostatic forces'.

'I think that the silver ions + chloride ions form together into a strong crystal lattice.'

But such responses were in a minority, and some of the answers reflected the alternative conceptual framework for ionic bonding discussed in Chapter 8, ie that ionic bonding was considered to result from electron transfer:

'The outer electron in the silver transfers from the outer shell of the silver to the outer shell of the chlorine. This is called ionic bonding.'

Students in post-16 groups tend to have a better mental image of the particles present in the solutions (although some still suggest sodium chloride solution contains sodium chloride molecules and not ions, or hydrogen and hydroxyl – or oxide – ions instead of water molecules), and are less likely to give the more obscure types of explanations found among younger students. However, they are still likely to suggest that the formation of the ionic bond in the reaction mixture is due to electron transfer:

RS•C

'The silver ion donates an electron to the chlorine ion, and an ionic bond, complete transfer of electrons, is formed.'

'The electrons from the silver transfer to the chloride creating a electrostatic attraction between the 2 atoms.'

The idea that an ionic bond is necessarily the outcome of an electron transfer event (see Chapter 8) may be so ingrained that a student will explain the formation of the bond in silver chloride in this way despite demonstrating a clear appreciation of the species present in the reaction mixture. For example, one student in a class of 16–17 year olds correctly identified the species in the sodium chloride solution as Cl^- ions, Na^+ ions and water molecules, and in the silver nitrate solution as Ag^+ ions, NO_3^- ions and water molecules. This same student recognised that after the silver chloride was precipitated the solution would contain Na^+ ions, NO_3^- ions and water molecules. Yet the formation of the ionic bond was explained in terms of atoms (cf Chapter 6) and electron transfer:

'The silver and chlorine atoms, lose and gain an electron, respectively. To form charged ions (Ag^+ and Cl^-), positive and negative. This results in an electrostatic attraction between the Ag^+ ion and Cl^- ion to form an ionic bond.'

It not surprising, given this view of bond formation, that this student thought that each ion was only bonded to one counter-ion (see Chapter 8) in the precipitate:

'one chlorine ion to each silver ion. This is because the silver ion has a valency of one. Silver ion is +1 and it needs -1 to balance charge chloride ion is -1.'

As well as this probe, the materials on **Precipitation** also include a study task which takes students through this reaction step by step and includes a sequence of diagrams (*eg* see Figure 9.7) to help students visualise what is actually occurring at the level of molecules and ions in the precipitation process.

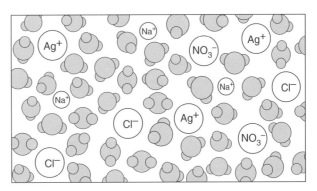

Figure 9.7 A diagram from the students' exercise on precipitation

How do reactions occur at the electronic level?

Often in post-16 level courses students are introduced to the idea of reaction mechanisms. Such mechanisms are very abstract (especially bearing in mind what has been reported above about some students having difficulty recognising the species present in reaction mixtures). Further, the common descriptions of reaction mechanisms introduce new formalisms regarding the representation of movements of single electrons and electron pairs (*ie*, so called 'curly arrows and fish-hooks').

In predicting reactions mechanism students need a quite sophisticated understanding of the patterns of electron density in molecules. Yet before they hope to apply such knowledge, they have to master the way such processes are formally represented.

One of the resources included in the companion volume is a diagnostic probe, **Reaction mechanisms**, comprising a pair of questions to test whether students are able to apply the formalism. Each question

RS•C

provides the student with the first step of a reaction mechanism, and asks the student to select the correct diagram from the options given (and to explain how they made the decision). The first question concerns an addition reaction, where the mechanism involves the movement of pairs of electrons. The first step is shown in Figure 9.8.

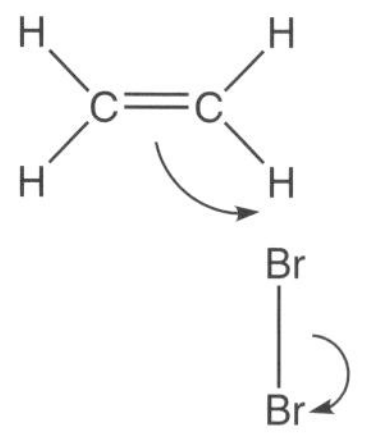

Figure 9.8 The movement of electron pairs

To predict the outcome of this stage, the student must appreciate where bonds are removed and formed, and the net 'gains' and 'losses' of electrons at each atomic centre (*ie* one carbon centre loses a 'share' in a bonding pair, one bromine centre gains fully an electron pair which it had only had a 'share' in).

The other question concerns a free radical mechanism, where electrons are considered to move independently and not in pairs (see Figure 9.9).

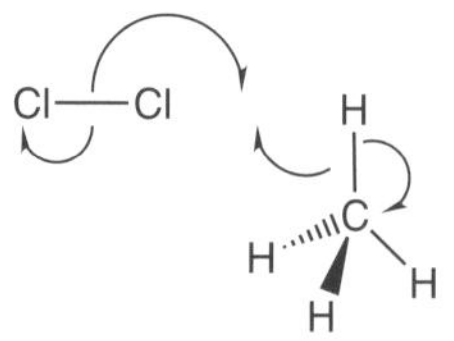

Figure 9.9 The formation of radicals

Again the student needs to appreciate where bonds are lost or formed. In this type of mechanism the atomic centres do not undergo net losses or gains of electrons (although, of course, there will be shifts in electron density patterns), but the students are expected to recognise any species which have unpaired electrons (at the carbon centre of the methyl radical formed, and one of the chlorine atomic centres).

When this probe was piloted for this publication some students were able to identify the correct diagram showing the next stage of the reaction, and to explain their choice.

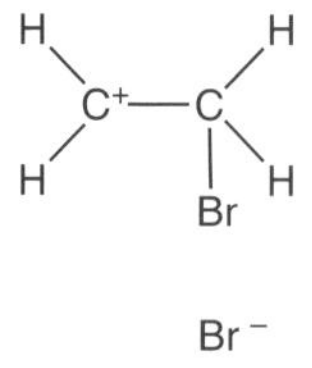

Figure 9.10 An intermediate stage in an ionic reaction mechanism

For example the correct diagram to follow the reaction step in Figure 9.8 is shown in Figure 9.10. Some students are able to select this answer, and explain their choice in terms of electron movements,

'because the double bond breaks and gives a pair of its electrons to one Br atom forming a bond. The other carbon from the double bond thus becomes positive. The electrons from the bond in the bromine molecule go to the 2nd Br atom therefore making it a -ve ion.'

RS•C

However, selecting the correct response from among the eight offered did not necessarily imply a clear explanation,

'The electrons have moved towards the left [sic] carbon atom from the double bond which then moved through the first Br to the second. This results in the C atom on the right being δ+ and the free Br ions being δ–.'

Some students select the wrong answer, and their explanations may demonstrate a limited understanding of the mechanism and its representation.

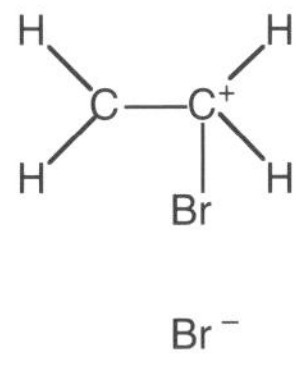

Figure 9.11 An alternative intermediate stage

So, for example, a student who selected the diagram reproduced as Figure 9.11, explained that

'because a plus charge is on the carbon as a result the differing electronegativity between bromine + carbon. The other bromine would gain a pair of electrons from the C-C double bond so would then be a separate ion.'

This response does not seem to call upon the electron movements shown in the question.

Some of the students who make the correct response seemed to be relying on recall rather than any understanding of the formalism. This approach is somewhat treacherous when the wrong responses look so similar to the correct ones (eg Figure 9.11 *cf* Figure 9.10, and Figure 9.13 *cf* Figure 9.12). So another student who incorrectly selected Figure 9.11 explained that,

'I remembered that the Br–Br bond separates & breaks the C=C bond to give the C a +ve charge. The bromine also becomes -ve charged when it is broken.'

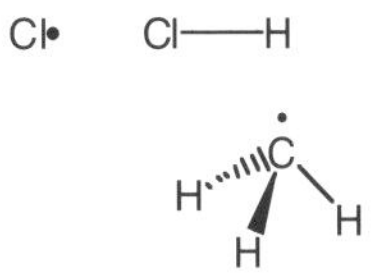

Figure 9.12 An intermediate stage in a free radical reaction mechanism

Figure 9.12 shows the intermediate stage that was the correct response for the second reaction mechanism (*ie* Figure 9.9). Again some students were able to select this option, and make fair attempts to explain their choice in terms of the electron movements,

'The Cl–Cl bond breaks, with one electron going to the left Cl, and one forming H-Cl bond. One electron leaves the H-C bond and completes the H-Cl bond, whilst the other electron moves to the C – leaving a Cl free radical, a HCl, and a CH_3 free radical'

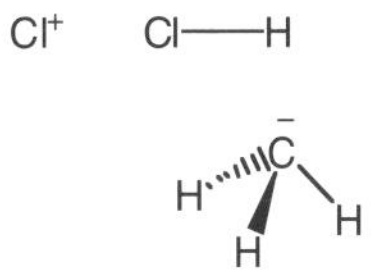

Figure 9.13 An alternative intermediate stage

RS•C

As with the ionic mechanism, some wrong responses demonstrated poor understanding of the formalism. For example, a student selecting the diagram reproduced here as Figure 9.13 was not paying heed to the meaning of the 'fish hook' arrows,

'The bond between Cl will break form[ing] Cl^- and Cl^+. one of the bond between C and H will break to form CH_3 and H^+. then H^+ and Cl^- will react to form HCl then leave Cl^+ and CH_3^-.'

Again, learning the mechanism by rote is not a fail-safe option. Another student wrongly selecting Figure 9.13 explained 'I have learnt this'.

Notes and references for Chapter 9

1. See the comments on chemical and physical change in Chapter 6.

2. As all reactions are technically reversible it has been suggested that the usual way of introducing reactions as unidirectional may be inappropriate (see note 3). If students adopt the implication that reactions 'go one way' then this could act as an impediment to later learning (see Chapter 4).

3. H-J. Schmidt, Should chemistry lessons be more intellectually challenging? *Chemistry Education: Research and Practice in Europe*, 2000, **1** (1), 17–26, available at **http://www.uoi.gr/cerp/** or **http://www.rsc.org/Education/CERP/index.asp** (accessed September 2005).

4. R. Peters, An Introduction to Chemical Equations, available at **http://www.wissensdrang.com/aufce12.htm** (accessed September 2005).

5. A government funded project to research and disseminate good practice in teaching chemical equations, the GENIUS (Giving Equations New and Intentional Understandings) project, has been based at the University of Reading, directed by Dr. John Oversby.

6. It has been argued by Alan Goodwin, of Manchester Metropolitan University, that chemical symbols and formulae may still be successfully introduced, and accepted by students, as a scientific way of representing substances and reactions. In this approach the two types of equation may be used together, with each acting as a 'cue' for the other. Students would learn that $CuSO_4$ is another representation for copper sulfate, without initially being asked to consider why that particular formulae is used.

7. See, for example, the data from the English National Tests, published by the Qualification and Curriculum Authority on the TestBase CD-ROMs, eg QCA (2001) TestBase 2000.

8. H-J. Schmidt, In the maze of chemical nomenclature – how students name oxo salts, *International Journal of Science Education*, 2000, **22** (3), 253–264.

9. It is known that many 14–15 year olds have not developed an adequate scientific model of combustion as a chemical reaction of a substance with oxygen (note 10). Indeed, it has been suggested that 'seen from the perspective of the learner, the demands are so great that combustion must be regarded as one of the last things we should expect our pupils to understand' (note 11).

10. R. Watson, T. Prieto & J. S. Dillon, The effect of practical work on students' understanding of combustion, *Journal of Research in Science Teaching*, 1995, **32** (5), 487–502.

11. P. Johnson, Children's understanding of substances, part 2: explaining chemical change, *International Journal of Science Education*, (in press).

12. K. S. Taber, An alternative conceptual framework from chemistry education, *International Journal of Science Education*, 1998, **20** (5), 597–608.

13. The figures given relate to reaction at 298K and are from J. G. Stark and H. G. Wallace, *Chemistry Data Book, 2nd Edition in SI*, London: John Murray, 1983.

RS•C

10. Constructing chemical conceptions

This final chapter reviews the ideas presented in earlier chapters and discusses how these ideas may be used to help students construct meaningful and acceptable versions of scientific concepts. The limitations of many student texts are considered, and some specific vignettes of classroom learning are used to illustrate key principles about teaching and learning. In conclusion the principles of constructivist teaching are reviewed.

Principles of constructing knowledge in the classroom

This publication has been written from a perspective – based on research into how learning occurs – known as 'constructivism'.[1] This approach can be summarised in a few simple principles:

1. People naturally, and actively, learn from their experiences, including, but by no means exclusively, experience in the classroom.

2. Our brains are not equipped to take on board large amounts of learning wholesale, as we have to process information in a limited 'mental scratch-pad'.

3. Therefore we need to break down information into manageable chunks that do not exceed the student's processing capacity...

4. ...before being later 'reassembled' into useful knowledge (and thus the references to constructing knowledge).

5. Meaningful learning occurs when we can make sense of new information in terms of what we already know.

6. Therefore the 'meaning' of new information is heavily determined by prior learning.

7. Students have already acquired a lot of ideas about scientific topics from their own practical experiences, and their interpretation of what they have been told, before they study those topics in school. This prior knowledge acts as the foundation on which new learning is constructed.

An experienced teacher comes to class with a vast storehouse of relevant ideas and knowledge relating to the topic being covered. In particular the teacher has an overview of the wider conceptual framework within which the ideas fit.

The student usually arrives in the same class with a much more limited and incoherent knowledge base about the topic, having conceptual frameworks that may include key omissions, major fractures and intrusions of alternative notions from various sources (see Chapter 4). Each individual student brings their own set of associations for words to class (see Chapter 3), and makes their own interpretation of what the teacher says (*eg* Figure 10.1).

RS•C

Figure 10.1 Students' interpretations and associations may vary
Modified with permission from the Journal of Chemical Education, Vol. 64, No. 9, 1987, pp.
766–770; copyright © 1987, Division of Chemical Education, Inc.

It is not surprising that what seems clear, simple and logical to the teacher may sometimes seem confused, complex and arbitrary to the student. Once we learn to recognise a pattern as a gestalt (of an atom, of an equation for neutralisation, *etc*) it requires a deliberate effort to decompose it into its constituent parts (see Figure 4.1). A gestalt is an organised whole in which each part affects every other part, the whole being more than the parts. It is difficult for the teacher to see the subject matter at the resolution available to the student.

What is more surprising, perhaps, is how sometimes students may continue to make sense of the teacher's ideas despite the most fundamental 'misconceptions'. As an example, consider the case of Annie, who held an alternative conception of ionic charges.

Vignette 1: The case of the deviant charges

When Annie enrolled on a post-16 chemistry course she already knew about ionic charges from her study of school science. Annie knew, for example, that the sodium ion was shown as '+1' , Na^+, and the ion of chlorine as '-1', Cl^-. However, Annie's interpretation of these charges was unusual. For Annie, the signs indicated deviations from a full shell electronic structure.

For Annie, Na^+ had a +1 charge because it had electron configuration 2.8.1. It had one electron over a full shell, indicated by the +. In a similar way, Annie thought that Cl^- had a configuration of 2.8.7, as the '-' indicated (to Annie) that this species was one electron short of a full shell. During nearly two years of post-16 chemistry Annie managed to interpret the teaching, and the comments of her classmates, in terms of her deviation charges.

Annie managed to make sense of ionic bonding from her own perspective, and was even able to suggest balanced ionic formulae – although not necessarily balanced in conventional terms. Her suggested stoichiometry for aluminium sulfate was $Al_4(SO_4)_2$ because to get a neutral compound 'you'd have to use, say four aluminiums, and, two, sulphates'. This incorrect answer was not due to a

RS•C

miscalculation, as Annie was able to explain why $(Al^{3+})_4 (SO_4^{2-})_2$ should be neutral.

In terms of her 'deviation' charges each Al^{3+} ion had three extra available electrons, and each SO_4^{2-} ion was lacking two electrons. Four aluminium ions provided 12 electrons, and the two sulfate ions accepted four of these to make up octet structures. This would seem to leave eight 'extra' electrons, but from Annie's alternative scheme this could be ignored,

'That'd make eight. It would make eight, so it would be neutral. Anyway it would give you eight, eight plus. [Interviewer: Would that be neutral?] A neutral charge ... because it would become nought.... if you had eight plus it's like having eight minus, you don't really have that because you have your shell with all your electrons in it, which could be eight.'

In other words, as deviation charges indicate a deviation from an octet, then eight electrons – an octet of electrons – counted as neutral. Annie's electron arithmetic only had to take count of the remainder when counting in base 8.

Amazing as it may seem, Annie's alternative take on ionic charge went unrecognised during most of her time in college. Neither her teachers, nor she, realised that they were talking across each other. Had Annie not been interviewed in some depth about her understanding of key chemical ideas shortly before her external examinations, then her alternative conceptions would probably never have been diagnosed. Even then Annie found it difficult to switch to the conventional scheme as her alternative ideas were well established, and had made sense to her.

The importance of diagnosing learners' ideas

To the teacher the signs '+' and '–' relate to electrical charges that students would be expected to know about from their study of physics topics in school. However teachers cannot assume the expected prior learning will be in place. As was found in Chapter 7, students may either be ignorant of the physical principles, or may simply not bring them to mind and apply them in a chemistry context.

The notion of being a 'learning doctor' was introduced in Chapter 4. Annie's alternative interpretation of 'deviation' charges was not a common conception that a teacher might have specifically been prepared for. If Annie's knowledge had been effectively audited at the start of her post-16 course then her alternative conception may have been diagnosed, and 'treated' early in her course. Students in subsequent post-16 chemistry classes in Annie's college were asked to undertake a set of induction exercises, which elicited a wide range of null learning impediments ('gaps' in the expected knowledge) and alternative conceptions of the fundamental chemical ideas that these students should have mastered during their earlier schooling.[2]

The importance of formative assessment, testing to determine what learning objectives are yet to be achieved, is now being recognised as very important,[3] and teachers are being urged to adopt strategies 'to explore pupil's progress and to help pupil's learning'.[4]

'Missing' learning, and alternative conceptions can be major causes of learning difficulties for students (see Chapter 4). Auditing prior knowledge, and, in particular, diagnosing substantive learning impediments, are activities which can be cost-effective in terms of time and effort, and which can help avoid some of the frustration felt when students fail to leave a well planned and executed lesson with the intended understandings.

The classroom materials included in the companioin volume can help diagnose some of the common alternative conceptions in key chemical topics. However, the materials are necessarily limited. Anyone teaching chemistry to students in the 11–19 age range will hopefully find some of the materials relevant and useful – but it has not been possible to address the vast catalogue of alternative ideas reported in the literature,[5] nor even to cover all the chemical topics met during this age range.[6] (A large set of Concept Cartoons is available to initiate elicitation of student ideas and discussion in a wide range of science topics. Although aimed at the 7–14 age range, many of the Concept Cartoons are suitable for some older students as well.[7])

RS•C

Building on the available foundations: only connect...

As well as discussing classroom materials in a number of fundamental topics, the previous chapters have illustrated an approach to teaching chemistry which takes into account key ideas about learning: the importance of matching classroom presentations to the students' existing level of knowledge; the need to break down material into manageable steps and to see the chemistry content at the student's 'resolution'; the importance of using aspects of existing knowledge to 'anchor' new ideas; the need to check students' interpretations and their meanings for words; the need to 'make the unfamiliar familiar' by using analogies where both similarities and differences are explored; and being concerned to find the optimum level of simplification that allows understanding now, without oversimplifying ideas so that they become learning impediments in the future.

A key aspect of this approach is knowing your students: knowing about the limitations imposed by their cognitive apparatus (such as working memory) and by their existing conceptual frameworks. Learners' ideas about science can be 'wacky', inventive and thought-provoking. They can also act as significant learning impediments (Chapter 4). They are, however, the only frameworks of understanding through which the teacher's words can be interpreted by the student; the only raw material available for building new knowledge, and the only substrate to which new ideas can attach. Reconstruction may often seen desirable – but simply ignoring the student's existing frameworks of knowledge is not a feasible option. Learning is largely about making new connections to existing conceptual frameworks: developing and extending the learner's internal 'concept maps' (see Chapter 3) stored within the brain.[8,9,10]

Teachers' subject knowledge

So the teacher does not only have to have a good grasp of the science of chemistry, but also of the students, and of the subject pedagogy – the science of teaching the subject (see Figure 10.2).

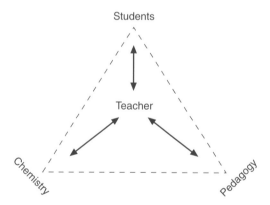

Figure 10.2 The effective teacher has to draw upon three distinct domains of knowledge [11]

Educational research, of the type that has informed this publication, can inform teachers' pedagogic knowledge. Such research leads to advice on ordering subject material, pace of presentation, likely problems with 'missing' prior knowledge or alternative interpretations and so forth. The teacher's own classroom experiences can also provide a wealth of insights in these areas.

Each class, and each student, is unique, so teachers need to be able to apply their pedagogic expertise in ways that are responsive to the needs of each learner – including students like Annie who come to class with ideas that are idiosyncratic. Published probes may not be available that will elicit such unique ideas, and teachers need to have honed their diagnostic skills as a learning-doctor (Chapter 4).

The third domain of knowledge that is important for the teacher is, of course, the subject knowledge itself. Although most teachers are highly informed about many aspects of their subject, it would be unrealistic to expect all science teachers to be experts in all areas of the science curriculum. Recently trained teachers in the UK are officially expected to demonstrate such expertise,[12] but even here one would expect some imperfections! Teachers, of all people, should value learning, and see themselves as life-long learners, with scope for developing their expertise.[13] Given that it is inevitable that teachers will have their own alternative conceptions and knowledge 'blind-spots', it seems sensible to encourage teachers to explore and develop their knowledge, rather than create an assumption of omniscience where practitioners can not openly admit to having an imperfect knowledge base.[14]

I remember having a disagreement with the Head of Chemistry in my first job. The topic of strong acids arose, and my colleague mentioned that strong acids had a pH value of 1. When I suggested that this was not always the case, he disagreed. I argued that strong acids could have a pH of less than 1 (which did not show up on the indicator paper used in the school laboratories), and that when diluted sufficiently the pH would rise above 1. This was clearly seen as a rather heretical suggestion. However, being reasonable people, we met in the laboratory at lunchtime and undertook sequential dilutions of a sample of bench acid until there was no doubt that the pH value was rising well above 1. The point is not that a teacher had an alternative conception, but that he was prepared to learn.

Of course there is a difference between teachers actually holding alternative conceptions and just having slightly distinct perspectives on a topic. In Chapter 2 we found that teachers could not agree on the accuracy of definitions of some very basic chemical terms – but this need not imply that some of these teachers did not know what is meant by the terms molecule, element and so forth.

In Chapter 3 'the' structure of chemistry was discussed, but just as students' conceptual frameworks will not match the curriculum version of the subject, individual teachers will hold in their cognitive structures unique versions of 'chemistry', each idiosyncratic and imperfect sub-sets of the formal structure of the subject. The teachers' mental concepts maps will be much more sophisticated, extensive and reliable than those of the students – but hardly encyclopaedic.

Learners' common alternative conceptions about chemical bonding were considered in Chapter 8. One key point made there is that students who learn about chemical bonding as being either ionic or covalent may find it difficult to later accept any other classes of bond. Some teachers may have sympathy with the students' views. Indeed an established and respected science educator has suggested to me that he did

'not really feel that metallic bonds are first rate chemical bonds – more the case of groups of metal atoms 'making the best of a bad job'. Hydrogen-bonds are certainly 'inferior' and I have some sympathy with the student who dismisses them as 'just a force, and not a real chemical bond' '.

However, the teacher's developed view is that there are graduations of bonding – where the student's less sophisticated approach is likely to be that examples that do obviously meet the criterion of filling electron shells are simply not bonds.

Of course, the formal structure of chemistry is an abstraction that does not exist in any one place. If teachers' versions are imperfect, then most practising chemists are likely to hold even less satisfactory versions of the subject – up-to-date and detailed in their immediate field, but often quite limited in areas of the subject that they have not had reason to think about for years or even decades. If 'the' current structure of chemistry can be said to exist anywhere, it is in the research literature of the subject – but few teachers can update their knowledge from the primary sources.

Textbooks as flawed authorities

Most teachers are likely to use student textbooks as their sources for checking information. Such textbooks are often written by practising teachers or by those who have taught, and are used by teachers to gauge the level of presentation needed. In Chapter 4 it was suggested that some substantive learning impediments (students' alternative conceptions and alternative frameworks) may be labelled as 'pedagogic'. In simple terms, some alternative ideas which block intended learning actually derive from the teaching of the subject itself.

RS•C

Many of the alternative conceptions discussed in this publication can either be found in student textbooks, or can at least be understood to be encouraged by such texts. A relatively cursory examination of a range of recently published books quickly revealed many examples of incorrect, dubious or unhelpful presentations.

Unhelpful descriptions of macroscopic phenomena

Students often have difficulty with the notion of acid strength because they reasonably assume that strong means concentrated (see Chapter 2). One textbook attempts to clarify the topic with the following introduction:

'Acid strength is different to how strong or weak an acid is; it is a measure of how much acid a solution contains, not how strong the actual acid is.'[15]

It is hard to see how this statement helps any learner make sense of the chemical distinction.

Another area of student difficulty is learning about chemical equations (see Chapter 9). Often equations given in student books do not clearly distinguish the material aspects and the non-material ones, *ie* energy is included in an equation using the same type face used for the substances. To give just one example,

fuel + oxygen → oxides + heat + light

In this case the oxides, heat and light were collectively described beneath the equation as 'the things [sic] that we end up with'.[16] As it is known that students do not always distinguish between matter and energy terms (see Chapter 6), this must be considered an unhelpful way of representing the chemical process.

Another book tells readers that 'a convenient way of writing down what happens in the reaction between zinc and dilute hydrochloric acid' is

'**zinc** + hydrochloric acid → **zinc chloride** + hydrogen'[17]

There is no explanation (or obvious rationale) for the decision to emphasise some parts of this equation in bold type. The term 'salt' is printed in bold type further down the page, so this equation could have been meant to be read:

zinc + hydrochloric acid → **zinc chloride** + hydrogen

This interpretation is pure conjecture, and, even if correct, students could hardly be expected to realise this. Clearly the use of such texts requires careful support from the class teacher.

Unhelpful descriptions of the molecular world

If textbook treatment of macroscopic phenomena may be unhelpful, attempts to help students learn about particle models may be even more flawed. It is known that this is an area where students often have difficulty (see Chapter 6), and so clear and accessible texts would be helpful.

Definitions of basic chemical concepts may be problematic, but some books for students seem to offer definitions produced with little thought. Many examples were given in Chapter 2, but further examples can be found in more recently published texts. So one book describes the atom as 'the smallest particle in an element'[18], a description that might be seen to be just as applicable to a molecule, or an electron.

One book aimed at 12–13 year olds defines element in terms of 'only one type of particle' and compounds in terms of 'more than one type of particle', which would be appropriate for ionic compounds but not covalent compounds. Further on in the book 'compound' is re-defined as 'two or more types of atoms chemically joined together to make a single type of molecule (particle)'.[19] As well as appearing contradictory, this approach does not clearly distinguish the macroscopic and molecular levels. The transition is obvious to author, and teacher, but is not made explicit to the student reader. (Further examples of this type of sloppy approach are discussed below.)

RS•C

Diagrams which are meant to clarify ideas, can potentially be just as confusing as carelessly written text. One book shows particles in a solid crammed together, and uses this model to explain properties of a solid. The diagram is repeated in a section called 'summing up' which reiterates that 'solids are made up of particles that are very close together'. On the facing page however, are two diagrams showing particles in a solid when it is heated. The text informs the reader that 'the particles begin to move further apart making the material bigger'. There is no discernible difference in the spacing of the particles before and after heating (which is realistic), just an attempt to show more vigorous vibrations about the lattice positions. However the particles in both figures are separated by significant spacing, so they are shown much further apart than the particles in previous diagrams showing solids and further apart than particles in diagrams representing liquids. Immediately following the diagram the text continues: 'The expansion of solids can cause a problem for designers.' It transpires this is a reference to designing bridges, but could equally apply to the graphic designers who prepared the figures.[20] The teacher will understand that the two different types of diagram of particles in a solid are used to make different points, but seen from the resolution of the learner, with a limited appreciation of the role of models in science (see Chapter 6), this contradiction must be very confusing.

It is not unusual for diagrams showing the relative particle separations in the three states of matter to represent the gas particles much too close together – sometimes separated by distances of only 1-3 particle diameters.[21] One book has a diagram showing that 'air is a mixture of different gases' which represents molecules of nitrogen, oxygen, carbon dioxide and argon crammed together as if in a liquid – with most molecular separations being less than one atomic diameter.[22] Another book has a similar picture showing 'air is a mixture of gases' with only marginally better spacing.[23] Perhaps the worst example examined was in a book for post-16 students, where diagrams showing 'the arrangement of particles in solids, liquids and gases' and 'the changes of state' showed the particles furthest apart in the solid and closest together in the gas – where, unlike in the condensed phases, two of the particles are shown to actually be in contact.[24]

Sometimes the molecular level explanations given in students' books must seem very obscure to learners. In one book a figure illustrating the difference between a concentrated and a dilute acid solution had the following legend:

'More particles of acid collide with the marble in a concentrated solution than in a dilute solution. The fewer particles of acid must now [sic] move past particles of water to get to the marble.'[25]

Perhaps this made sense to the author, but surely students are left to fill in too much information to attempt to understand this explanation.

A textbook spread about diluting solutions in one book uses orange squash as its example. Although some students might not initially realise that this is a mixture, the text refers to how 'the colouring and flavouring get spread out'. Later on the page students are asked to 'imagine that the orange drink is made of orange particles'. A diagram shows 'a really dilute solution with only one orange particle left in it'.[26] Although this is presented as a thought experiment, it does show the mixture as represented by a single type of solute particle which could be confusing. The text also suggests that 'the orange [squash] in the bottle contains a lot of orange particles, and not many water particles' which seems very unlikely for a solution. Moreover, the diagram shows particles so large that only five fit across the beaker. The ratio of orange particles to water particles is 1:50, which a 'back-of-the-envelope' calculation suggests is equivalent to a concentration of about 1 mol dm^{-3}. In the context of school science this would hardly be a 'really dilute solution'.

What is not clear in this particular case is whether the diagram showing 51 particles of the solution filling up the beaker is intended to represent the particles in the solution, or a macroscopic scale model. Teachers and authors are able to use a wide variety of modelling conventions, but as it is known that most students have naive ideas about modelling in science (see Chapter 6), it is important that the conventions used in such diagrams are made explicit for the learners.

RS•C

A diagram of glucose solution in another book shows the water molecules as having three separate spheres joined, but the glucose molecules as hexagonal structures the same size as the water molecules.[27] It is not clear why this type of representation was chosen, but if students know that the shape of the water molecules is meant to reflect the 'H_2O' structure with three atomic centres, then sharp corners on the same overall sized glucose molecules would seem to imply that they cannot be composed of the same sort of components as water molecules.

A common error in many books, that is often reflected by students, is that the third and subsequent electron shells, like the second shell, can only hold 8 electrons.

'Electrons are arranged around the nucleus in shells. Each shell can only hold so many electrons. The first shell can hold up to 2 electrons. The second and third can both hold up to 8 electrons.'[28]

The logic behind the 2, 8, 18, 32... pattern is too abstract to be presented in school science. However, the simple notion that the larger the shell, the more (mutually repelling) electrons it can accommodate would seem accessible.

In one text for post-16 students 'covalent bonding in an iodine molecule' was represented as two overlapping circles with electrons as dots and crosses. The legend reported that 'only the electrons in the highest energy level [are] shown.' However, the accompanying diagram showed the 14 valency ('outer shell') electrons in the iodine molecule, *ie*, the six non-bonding ('lone') pairs and the bonding pair. The bonding electrons would, of course, be in a molecular orbital at a lower energy level than the electrons in effectively unperturbed sp^3 atomic hybrid orbitals.[29] This is not a pedantic point when it is remembered that if the bonding and non-bonding electrons were at the same energy level the molecule would spontaneously dissociate. A diagram meant to help explain bonding actually undermines the physical basis of that very phenomena. Research shows that post-16 students may have difficulty distinguishing the concepts of electron shell, sub-shell, orbital and energy level (see Chapter 7) and inaccurate textbooks are only likely to exacerbate this.

Confusing the macroscopic and the molecular

The molecular model is extremely important in chemistry, although it is an abstract model that many students find difficult (see Chapter 6). Many explanations in chemistry require transitions between the macroscopic and molecular levels, when it is believed that such transitions place high cognitive demands on students.[30] Moreover, it is known that students often mis-apply the molecular model by ascribing macroscopic properties to the particles at the molecular level. This means that it is important for teachers to clearly distinguish between these two levels, and to carefully highlight when there are transitions between them. The same onus falls upon textbook authors. The necessary care is not always taken, so that student books include headings such as 'formation of ions from elements',[31] and comments such as 'a formula shows the number of atoms of each element found in a compound'[32].

Diagrams often confuse the two levels, sometimes apparently deliberately (as in the example of diluting orange squash discussed above). A diagram showing a metal sheet being rolled shows the number of particles in a thickness of metal before and after being milled. This is perhaps an attempt to emphasise that the size of individual particles stays the same, but students may not appreciate the schematic nature of the sheet thickness being reduced from 8–9 atoms to 2–3 atoms.[33]

It is more difficult to appreciate the rationale behind a diagram of 'smoke particles [*sic*] bombarded by many air molecules' which shows one particularly irregularly shaped object, much like an asteroid in appearance, and seven smaller spheres. In its longest direction the 'smoke particle' has a length equivalent to four times the diameter of the 'air molecules'. Not only are the relative sizes completely wrong, but the highly complex shape of the smoke particle implies that it has a detailed structure at a scale much smaller than that of air molecules.[34]

A diagram of a solution in one book shows the solvent molecules as spheres, but shows the solute particles as cubes, much like tiny salt or sugar grains. There is a coloured background between the

RS•C

particles in the solution (which could encourage the alternative conception that particles are embedded in 'substance'). In a diagram showing filtration of a suspension, the suspended particles are shown as if macroscopic. A few pages later in the same student book a diagram of salt water solution has a flask of blue liquid with small, but visible cubes mixed into it.[35] As one final example, a diagram showing the action of enzymes in one book represents the enzyme as molecule-sized scissors (see the discussion of classroom analogies below).[36]

Encouraging an atomic ontology

One particular alternative conception that was highlighted in Chapter 6 was the 'assumption of initial atomicity' – the idea that all chemical process occur between atoms. Yet many textbooks present their expositions as if chemistry does start with discrete atoms. Consider the following definitions,

'[atom] A particle of an element that can take part in a chemical reaction.'

'[compound] A substance made from the atoms of two or more elements that have joined together by taking part in a chemical reaction.'[37]

Books may ask students 'how many chlorine atoms [sic] react with a single atom of magnesium?'[38], and commonly present diagrams representing reactions in terms of isolated atoms, even when one or more molecules worth of atoms are needed (four atoms of hydrogen with one of carbon; two atoms of oxygen with one of carbon).[39,40,41,42,43]

This common type of misleading diagram may have been perpetuated because authors do not closely think about the actual process they are trying to illustrate, or perhaps there may be a deliberate attempt to simplify the real chemical reactions. Perhaps there is a third possibility – that sometimes the authors genuinely believe the explanations are valid (see below). Whatever the reason, such an approach conveniently allows authors to imply that there is a simple reason why reactions occur,

'When two non-metals such as chlorine and hydrogen react, they do it by sharing electrons. The diagram shows what happens to the shared electrons…A hydrogen atom has just 1 electron in its first energy level. A chlorine atom has 7 electrons in its third energy level. If the two atoms share 1 electron each…hydrogen can fill its first energy level and chlorine can fill its third energy level.'[44]

Although the statement above is a fair description of the interaction between two isolated atoms, it is hardly surprising if students assume it is meant to relate to the chemical reaction between (molecular) hydrogen and chlorine. These types of statements often seem to include subtle 'sleight-of-hand' transitions between the macroscopic and molecular levels, during which discussion of isolated atoms is suddenly applied to the macroscopic substance (which are not atomic). Some textbooks explicitly suggest real chemical reactions can be explained in this way,

'How chemical reactions work:
All atoms, apart from the noble gases, will form compounds. When they form compounds, it's all to do with electrons. They try [sic] to make a full shell of electrons on the outside of the atoms.'[45]

Note that in this example the narrative slips from the molecular level (atoms) to the macroscopic (noble gases, compounds), and then back (electrons, atoms).

'Why elements react to form compounds:
Atoms like [sic] to have each energy level either completely full or completely empty just like they are in the noble gases. The atoms are then more stable. This is why sodium reacts with chlorine.'[46]

This statement is made so definitively that the student reader would not be expected to be expected to realise that the discussion of what 'atoms like' has no relevance to the reaction of (metallic) sodium and (molecular) chlorine. One wonders whether the authors of this statement may themselves actually believe this is a valid explanation.

The responses commonly obtained from the **Hydrogen fluoride** probe (see Chapter 9) are hardly surprising if students have been exposed to such texts. Not only is the implication that molecular materials react because of the unstable nature of atoms, but anthropomorphic language is used.

RS•C

In the following, final example, the rationale for reaction is related to energetics rather than some irrelevant consideration of the electronic structures of isolated atoms. Yet the notion of atoms reacting is still introduced (for no clear reason), and they are said to react with oxygen (not particles of oxygen).

'Fuels are made of molecules. The atoms [sic] that make up these molecules react with oxygen [sic]. The new molecules made in the chemical change contain less stored energy than the fuel molecules. This is because [sic] some energy is released.'[47]

Despite the encouraging attempt to give a scientifically valid rationale for the reaction occurring, the final sentence manages to reverse cause and effect, suggesting that the release of energy causes the products to be at a lower energy level than the reactants. The release of energy is actually a consequence of this difference in energy levels.

The example of ionic bonding

In Chapter 1 the octet framework was used as an example of a common alternative conceptual framework found among secondary school and college students. At that point it may have seemed inexplicable that so many students should build up a similar set of invalid ideas about central aspects of chemistry. Yet it is clear that many of the statements and diagrams in students' textbooks encourage learners to think in terms of chemical reactions which occur between unbound atoms that are actively seeking to fill their shells by forming chemical bonds. This alternative conceptual framework is surely an example of a pedagogic learning impediment (Chapter 4) – something that largely derives from the way the subject is taught.

Before moving on, it is useful to illustrate this point with the more specific example of the molecular framework for understanding ionic bonding (see Chapter 8) which may be considered to form a part of the wider octet framework. It has been found that students commonly equate ionic bonding with electron transfer between atoms to form ions: the bond is considered to only exist between the particular ions formed by specific electron-transfer events, so in sodium chloride each ion is only bonded to one other, and the resulting ion-pairs are seen as having molecule-like status (see Figure 8.2).

Students' common alternative conceptions of ionic bonding can be found to be well represented in some books. One book examined even presented a diagram which 'shows the molecules [sic] in seawater'. The diagram showed five molecules of water mixed with five 'molecules' of sodium chloride NaCl.[48] Perhaps at such an incredibly high concentration (about three parts salt to one part water, by mass!) the ions would associate, but this is certainly not a reasonable image of the particles present in seawater.

Another book suggests that 'salt is made from [sic] the elements sodium and chlorine. There is one atom of sodium and one atom of chlorine.'[49] In this short extract there is an implication that the salt that students will be familiar with from the dinner table would have been manufactured from the elements, and then a transition from the macroscopic level to the molecular level, where reference is made to single atoms, implying the elements are atomic. On the following page a diagram of a 'salt molecule' shows one smiling sodium atom holding hands with one frowning chlorine atom.[50]

Another books reports how

'It is unlikely that a single sodium ion and a single chloride ion would ever find themselves alone together! But it is possible, and this would be the smallest part of the compound which still has the properties of sodium chloride.'[51]

Clearly, most of the properties of sodium chloride would not be shared by a single ion-pair. The formation of an ionic compound (such as silver chloride, see Chapter 9) does not require electron transfer. Yet books for students make statements about 'the two basic principles underlying the formation of bonds: electron transfer and electron sharing'.[52]

RS•C

Diagrams that purport to be showing 'the reaction between lithium and fluorine' or 'the formation of ionic bonds in sodium chloride, magnesium oxide and calcium chloride', or 'bonding in calcium fluoride' actually show electron transfer between isolated atoms.[53,54,55] Indeed these electron transfer diagrams are virtually ubiquitous in student texts, and (as in the case of reactions to form covalent compounds) atoms are usually drawn even when one or more molecule's worth of atoms are needed.[56,57,58,59,60,61,62,63] It has been found that even trainee chemistry teachers find the molecular model of ionic bonding as acceptable.[64] This can be understood if they are preparing their classes with many of the current textbooks, and it would not be surprising if they actually go on to teach these erroneous ideas. (Perhaps some of them will one day write student texts that perpetrate myths such as the assumption of initial atomicity, and the full shells explanatory principle!)

Periodicals such as Education in Chemistry and School Science Review often receive letters from readers disagreeing with the interpretations of science put forward by authors of articles (who are often also teachers). For example, the following extract is from a letter objecting to an article in the School Science Review which had criticised the notion of sodium chloride molecules. Seeing ion-pairs seen as molecules is a common alternative conception found among students (see Chapters 7 and 8), but the following view point was expressed (some years ago) by a teacher:

'Is it so wrong to refer to a sodium chloride molecule? In the crystal, every sodium ion can be paired off with a neighbouring chloride ion, even though the bonding forces holding them together may be electrostatic rather than covalent. Even in a solution the ratio of the numbers of the two ions is still 1:1, and thus even these we can think of NaCl as a sort of basic unit.'[65]

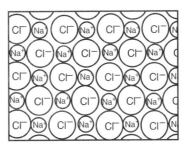

Figure 10.3 Solid sodium chloride

The comment that 'in the crystal, every sodium ion can be paired off with a neighbouring chloride ion' reflects research findings that students will, when shown a diagram such as Figure 10.3, tell the interview which (single) cation is bonded to which (single) anion. The ions can be 'paired off' – see Figure 10.4 – but this does not relate to the actual interactions in the lattice, nor to the properties exhibited by the ionic crystal!

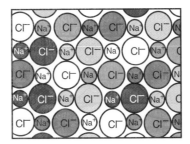

Figure 10.4 The way students sometimes conceptualise sodium chloride NaCl

When the same teacher commented that 'even in a solution the ratio of the numbers of the two ions is still 1:1, and thus even these we can think of NaCl as a sort of basic unit' he was making a valid point about stoichiometry, and perhaps implied the term 'basic unit' in the sense of the empirical formula. However, many students mean something rather different by references to NaCl molecules. It is important that students realise that the main species present in such solutions are the water

RS•C

molecules, and the hydrated ions (see Figure 10.5) and not – as many think – ionic molecules (see Chapter 8).

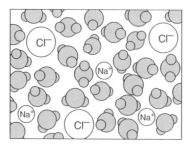

Figure 10.5 Sodium chloride solution

One major concern is when science teachers are teaching outside of their subject specialism. There are non-specialist guides which are designed to help non-chemists plan the teaching of chemistry topics, and these can sometimes be very useful.[66]

However, even these teachers' guides may contain errors. It was pointed out in Chapter 8 that students commonly associate ionic bonding with electron-transfer, which can encourage the development of major alternative conceptions. Yet one guide to teaching secondary chemistry advises teachers that:

'The essential idea for pupils to appreciate is that in ionic bonding metals transfer electrons to the non-metals such that the nearest noble-gas electronic structure is obtained for both the metal and the non-metal ... Appropriate pictorial representations are crucial in this regard.'[67]

The type of representations referred to are the very type of diagram (of electron transfer between isolated atoms) that research has suggested are so inappropriate (see Chapter 8, Table 8.2). The author's figure supposedly showing 'the formation of sodium chloride, shown in terms of the particles involved' has virtually no relevance to any likely chemical process for forming sodium chloride. However, examination questions are regularly set asking students to draw such figures.

Problems of changing conceptions: challenging versus developing students' ideas

Some alternative conceptions are readily overcome, but others are more tenacious. In Chapter 1 it was suggested that it is useful to distinguish between discrete alternative conceptions and alternative frameworks. An alternative framework is a coherent and integrated set of conceptions, some (but not necessarily all) of which are 'alternative' to scientific ideas. Because the ideas are connected, they are mutually supporting: the veracity of any one conception seems to be assured by its relationship to the whole structure. Our conceptual frameworks are judged (usually subconsciously) by their 'explanatory coherence'.[68] Sets of ideas that seem to fit and have generally worked well together are likely to be retained. Even a few 'shaky' conceptions may be considered acceptable in the context of a generally successful explanatory framework – at least until something with greater overall coherence is available. Perhaps this explains why some experienced teachers writing textbooks seem to adhere to the octet framework in their explanations. Perhaps these authors do think of chemical reactions in terms of atoms trying to fill their shells by sharing or transferring electrons!

Teachers can often find that despite offering students scientifically better ways of looking at a phenomena, the students soon revert to their prior alternative conceptions. This can be disappointing, as one of the teachers piloting materials included in the companion volume reported,

'I found the questions and the students' responses interesting. Even from a superficial glance at their answers, it is clear that the complete shell of electrons dominates their thinking. We have spent some time looking at stability in terms of energy changes and forces between charged particles, so it shows me how easily people revert to simple and familiar ideas.'

RS•C

A teacher may be proposing to help the student construct knowledge which is more scientific than existing conceptions, but if this means whole-scale demolition of familiar conceptual frameworks the student may tacitly deny the teacher 'planning permission'.

If learning can only take place in small steps, building on existing knowledge, and yet some alternative conceptions are protected by being integrated into self-supporting frameworks, then bringing about major conceptual restructuring may seem an unlikely target. In practice 'conceptual revolutions' do occur – as evidenced by the shifts in perception brought about by scientists such as Newton, Darwin, Einstein, Meitner and Lavoisier. Few students are likely to match the feats of such great scientists unaided, they need teachers to help scaffold their learning (see Chapter 5).

Lavoisier was astute enough to recognise where the existing conceptual structure of chemistry was challenged by anomalous data. He was able to explore a new set of ideas and interpretations, and compare them with the (then) current phlogiston theory. He was able to construct a new framework for chemistry that he ultimately realised explained the facts better. Students can be taken through a similar process.

Eliciting students' ideas is not an end in itself, but is the precursor to finding ways to demonstrate where those ideas are inadequate (through class practical work, demonstrations, calculations, thought experiments *etc*), and 'planting the seeds' of doubt.[69,70] Some of the classroom materials in the companion volume provide opportunities for students to become aware that their alternative conceptions do not match the facts. The intellectual dissatisfaction or 'cognitive dissonance'[71] or 'dis-equilibrium'[72] produced may well motivate students to find more satisfactory explanations – but the teacher should always remember that the student's judgement of what is satisfactory is made from the context of their wider conceptual frameworks, and not from the perspective of the teacher's knowledge base. It may take months, or even longer, for students to give up particularly well established and integrated frameworks. In time, these shifts can be brought about, as the example of a shifting conceptual profile in Chapter 8 (see Figure 8.20) demonstrates – but it may be a slow process. Clearly the teacher has to keep reinforcing the scientific view whenever suitable contexts arise.

Another theme that arises from the ideas discussed in this publication is the way that learning is an active process – that is, it requires activity on behalf of the learner and not just the teacher. This is needed at all levels of the educational system.[73,74] Students need to be able to explore and play with ideas. The key type of activity is mental,[75] and with some classes a great deal of active processing of ideas can occur with students sitting quietly in their seats, and the teacher scaffolding learning through a form of Socratic dialogue (promoting independent reflection and critical thinking). Teaching materials which use DARTs and provide scaffolding PLANKs and POLES will help ensure students are thinking, and not just copying notes for rote learning (see Chapter 5). However, with younger students in particular, the constructivist classroom can be a busy place, with students developing arguments and tests for their various ideas, and the teacher perhaps ceding more control than feels comfortable.[76]

It is sometimes implied that there is a tension between two 'opposing' views about responding to students' alternative conceptions. Learners' ideas may be seen as obstacles to be demolished and overcome, or as the conceptual resources that need to be developed to become more scientific. In practice deeply held conceptions are probably never completely 'forgotten' and so a simple substitution model of conceptual change is naive. Conversely 'development' must involve persuading students to build new conceptual frameworks which are organised differently, have some different components, and which ultimately will be used in place of the existing ideas. The 'challenge' versus 'develop' dichotomy is an artificial one. The building process can only use the foundations available, but ultimately looks to change students' minds.

Planning to avoid learning impediments

One of the themes met in previous chapters is the need to plan teaching so that students are introduced to ideas in a logical way that helps them construct knowledge. There are two main aspects to this planning. One involves analysing the content itself, to make sure that key concepts are

RS•C

introduced and understood before more complex ideas that build upon them. The second aspect is not to make undue assumptions about students' existing ideas.

Learning from pedagogic learning impediments

The purpose of labelling some ideas learners bring to class as pedagogic learning impediments is to remind us that if teaching has contributed to their development, we should be able to avoid them in future by changing our teaching. For example, it is known (see Chapter 7 and 8) that many students:

■ conceptualise ionic bonding in molecular terms;

■ conceptualise metallic bonding as covalent and/or ionic;

■ conceptualise giant covalent structures as consisting of discrete atoms or small molecules.

This can clearly be understood in terms of the order in which presentations are made. Usually covalent bonding in simple molecules is studied first, where valency determines bond number, and small discrete molecules are formed. When ionic bonding is discussed the student uses the prior knowledge of the covalent case to make sense of the new learning: so electrovalency is seen as determining the number of bonds formed, and ion-pairs are seen as molecules. (This is of course much more likely to occur when ionic bonding is incorrectly shown as electron transfer between isolated atoms!) The student now has two 'mental slots' (see Chapter 9) for bonding, and when metallic bonding is encountered it is often seen in terms of being covalent or ionic. Giant covalent structures are also seen as molecular (and intermolecular bonding in simple molecular lattices is often ignored).

A consideration of students' ideas such as these, and an analysis of the relative complexity of the different types of structure (see Chapter 3) leads to the suggestion that students are less likely to develop common alternative conceptions if the teaching order was changed. The following teaching order (see Table 10.1) suggests starting with the simplest case, and moving to more complex structures. According to this view, the 'simple covalent' case is actually the most complex, because it needs two types of bonding to explain structural integrity, and so it should be taught last.

Type of structure	Bonding	Comments
1. Metallic crystal	Metallic: cations (atomic cores) and delocalised electrons	One element present; charge on cation related to valency
2. Ionic crystal	Ionic: cations and anions	Added complication: two (or more) elements; stoichiometry determined by charge ratios
3. Giant covalent	Covalent	Added complications: number of bonds (and stoichiometry, if a compound) determined by valency; bonds have specific directions
4. Simple covalent	Covalent intramolecular, plus intermolecular (van der Waals, H-bond)	Added complication: additional level of structure – need to consider discrete molecules, and arrangement of molecules in crystal.

Table 10.1 A teaching order for solid structures[77]

RS•C

Vignette 2: Never assume…

The planning of teaching needs to allow for students' prior knowledge, but as has been shown throughout this publication this cannot be assumed to simply reflect the topics previously covered. Good advice to teachers is to always check prior learning , and not to make assumptions:

■ Do not assume that learners have an understanding of previous topics which matches the curriculum version of the science;

■ Do not assume that learners have no relevant ideas about topics which they have not previously been taught.

Experience suggests that one can add

■ Do not assume the students will already have covered something in another subject.

In Chapter 8 it was suggested that by the time most students complete their secondary education they will have specifically learnt about two types of chemical bonding (ionic and covalent). When they meet chemical bonds, and try to make sense of them, they will do so in terms of this available background knowledge. If a student sees a reference to a hydrogen bond, with no further explanation, it is likely to be interpreted within the existing conceptual framework The student has the two mental 'slots' (see Chapter 9) for making sense of bonds, and hydrogen bonds (involving hydrogen and another non-metal, usually represented by lines, no charges shown) are likely to be fitted neatly into the slot for covalent bonds.

For example, one post-16 student, Paminder, used the construct 'hydrogen bonding present' when she was undertaking an exercise discriminating between pictures of chemical systems (Kelly's triads – see Chapter 2). This was unexpected as she had not been taught about hydrogen bonds in chemistry at that stage of her course. She suggested that hydrogen bonds were present in a methane (CH_4) molecule. Paminder had acquired the category of hydrogen bond, but had subsumed it within her existing category of covalent bonds. She later explained how she had been introduced to hydrogen bonding in biology,

'at the moment we're doing like about DNA and double helixes, and DNA consists of like bases, of three things actually … But they've got bases, and they're joined by hydrogen bonding. But the hydrogen bonds are actually holding two bases together.'

However, when asked to explain what hydrogen bonding was she simply described a covalent bond involving hydrogen,

'say for example hydrogen gas, that consists of two atoms of hydrogen, and when they bond they, each one has one electron in its outermost shell, and when they bond, they bond like covalently. And that's what hydrogen bonding is. That's an example of hydrogen bonding.'

Perhaps Paminder's biology teacher incorrectly assumed that she would already have been taught about the concept of hydrogen bonds in her chemistry classes. One of the key messages of this publication is that teachers need to check students' prior learning. As the colloquial spelling aide-mémoire suggests: 'never assume – it makes an ass of 'U' and me'.

It seems unlikely that Paminder's case is unique. A recently published text for post-16 biology introduces the term 'hydrogen bonding' without explanation.[78] An accompanying diagram shows hydrogen bonds in two protein structures. In a schematic representation of the α-helix the hydrogen bond is shown as a dashed line between two parts of the structure. (In the second schematic, for the β-pleated sheet structure, the label for the hydrogen bond appears to be incorrectly pointing to what is presumably an amino acid residue.) A student meeting hydrogen bonding in this context is given no information to help make the learning meaningful. It is hardly surprising therefore that many just assume the hydrogen bond is a bond involving hydrogen. It seems reasonable for the learner to expect that If hydrogen bonding was a new and significant class of bonding, then this would be explained when first introduced.

RS•C

As most students enter post-16 courses with a strong commitment to the 'full shell explanatory principle' (see Chapter 8), *ie* erroneously believing that bonds form so that atoms may attain octet structures of full shells, the introduction of a new class of bonding that cannot be construed in these terms calls for careful exposition. If biology teachers take their lead from texts which just assume a knowledge of hydrogen bonding then there are likely to be many Paminders in post-16 classes who have formed the alternative conception that hydrogen bonding is a covalent bond to hydrogen, before anyone tries to explain the concept to them.

Making the unfamiliar familiar

Teaching can be seen as being about making the unfamiliar familiar. The obvious way to help a student become familiar with Beethoven's fifth symphony would be to play it to them, and the obvious way to get students familiar with paper chromatography would be to get them to try it for themselves. However direct experience is not really possible when the unfamiliar is the concept of oxidation numbers, or spin-pairing, or d-level splitting.

One technique commonly used by teachers is to use metaphor or analogy. These are ways of comparing the unfamiliar with something that is familiar, and so make it seem familiar itself. Metaphor is a key aspect of language: indeed it has been suggested that most of our language has derived from metaphor. So we have stories that are deep and those that are tall. People are stars and peaches. Many of our metaphors are 'dead'. This means that with repeated use they have become accepted as literal meaning (and a 'dead metaphor' is an example of a metaphor – the metaphor was never alive!) So we have long pauses – where a word used to describe a length is now accepted as describing the 'length' (duration) of a period of time.

It is important to remember that metaphors have 'hidden meaning'. When we say that there was a pregnant pause, we expect the listener to understand that a pause (however 'long') is not literally pregnant, but may be compared with a pregnancy in some way. Metaphorical language is poetic, and relies on the listener (or reader) to interpret the hidden meaning. When the reader is sophisticated, then metaphor can be very effective. A number of metaphors have been used in the text of this publication. The notion of a learning-doctor; references to conceptual structure as a substrate, to which anchors need to hook; to conceptual flotsam and jetsam; to icebergs of knowledge, and to 'gaps' in knowledge; to mental slots (and concept maps); to raw material for building learning (when planning permission is granted); and to seeds of doubt. Part of the rhetorical 'force' of such metaphors seems to derive from the very way that they are 'planted' in the text without explicit explanation.

However, metaphor may not be so effective when used with students who may lack sophisticated language skills, and who may expect teaching to be more literal. For example, social metaphors are often used to introduce students to the unfamiliar world of the chemist's molecular models. Sometimes this is done in quite unsubtle ways,

'The combining power of each atom is its valency. Think of the valency as the number of hands that each atom has to hold on to another atom.'[79]

Students seem to take to this notion of atoms as being like social beings, who enter into various relationships with other atoms. This helps them get an image of the molecular world, but sometimes they may have difficulty moving on beyond this stage.[80] For many students the atoms are actively trying to get full shells. If students are satisfied with this level of explanation, then there is no intellectual motivation to learn a more abstract explanation.

Using analogies to anchor to the conceptual bedrock

A main disadvantage of using a metaphor in teaching, is that is has 'hidden meaning' which needs to be 'unpacked' by the reader or listener. Similes are more suitable, as there is an explicit effort to make a comparison between the unfamiliar and existing knowledge. However, even when such a reference is made, it is necessary to spend time to make sure the comparison is explored. Consider an example from a student textbook, introducing the idea of atoms. An initial statement that 'Scientists have

RS•C

studied the behaviour of atoms since 400 BC...' seems rather dubious: certainly CERN has not been around quite that long. The book goes on to ask students to make the comparison between atoms, and bricks and stones. The notion of 'atoms as building blocks of matter' was criticised in Chapter 6, but putting such concerns in abeyance, consider how helpful the following extract is to the student just learning about atoms:

'There are about 112 different kinds of atoms. Each one kind is a separate element. [Note the invalid transition from the molecular level to the macroscopic] ...How can these tiny atoms be different? It helps to think of stones or bricks. Bricks do the same job (they join with others to make the wall), and they are made of the same kinds of substances. However, you can mix the ingredients in different ways and amounts to make different kinds of brick. Put these together and you get different kinds of walls.'[81]

Are the bricks meant to be the atoms, or the sub-atomic particles that make up the atoms and so make them different? Do bricks do the same job as atoms, or as stones? Are bricks made of the same kinds of substances as atoms, as stones or as one another?

I find myself at a loss to understand exactly what is being compared with what in this explanation. I find it even harder to see what a student is meant to make of this. It is no wonder that students learn ideas in science uncritically ('reactions occur for atoms to get full shells', see Chapter 9), when even attempts to make the unfamiliar familiar can be so confusing. One key point in using analogies is to be explicit about how they map onto the target concept.

Vignette 3: 'The MCG analogy is analogous to the Yankee Stadium analogy'...

In a fascinating paper about a students' learning about key chemical concepts there is a reference to a Melbourne Cricket Ground (MCG) analogy that was 'used in class' when teaching about the atom.[82] As the authors were reporting work undertaken in Australia it is likely that a comparison between an atom and the MCG would make sense to the students in the class. Presumably the size of the nucleus in an atom was being compared with a relatively small object (a cricket ball) placed in the middle of the MCG.

The paper was published in an international journal, based in the USA, and a note had been added to the end of the paragraph to explain that 'the MCG analogy is analogous to the Yankee Stadium analogy...'. One of the authors decided that the MCG example was too obscure for readers in some countries, and that the analogous analogy (presumably of a baseball at the centre of the Yankee stadium) would be more familiar. So here we have an example of an analogy made between two analogies. The original analogy between an atom and the Australian sports context was judged inaccessible to some readers, and so a further analogy was drawn.

For someone teaching in the UK the baseball example would not be particularly familiar to many students, and a better example might be from association football. For UK teachers, then, the journal perhaps could have explained that:

'The MCG analogy is analogous to the Yankee Stadium analogy [which is analogous to the Wembley Stadium analogy].'

Target	Nucleus	Atom
Analogy for Australian context	Cricket ball	Melbourne Cricket Ground
Analogy for USA context	Baseball	Yankee Stadium
Analogy for UK context	Soccer ball	Wembley Stadium

Table 10.2 An analogous set of analogies

RS•C

In other words, the reference to the Yankee Stadium was introduced to help the reader who was not familiar with the MCG: but for many readers in the UK, NZ, the Indian subcontinent, the West Indies and southern Africa, the baseball example would be less familiar than the original cricket reference. This is not intended as a criticism of the authors' selection of examples, but just a useful reminder that our judgements of what is and is not familiar to others can sometimes be mistaken.

In Chapter 7 the analogy between the atom and a solar system was considered. It was suggested there that this analogy derives from an assumption that students will be familiar with the structure of the solar system, and can use this knowledge to form an initial image of the atom. As was reported in Chapter 7, the assumption that learners are better informed about the solar system than the atom may not always be justified. Not only do analogies have to be explicitly mapped, they also have to be selected so that the analogue is genuinely familiar to the students.

Vignette 4: The parable of the molecular scissors

'Enzymes are biological scissors cutting up large molecules and making them into smaller more manageable pieces.'[83]

So part of the craft of the teacher consists of making the unfamiliar familiar by making comparisons with what is already known. Teachers therefore need to have the imagination to be able to think up useful analogies. The analogy will only be useful when the connection between analogue and target is clear, and when the analogue is genuinely more familiar than the target concept. A student who has seen the 'kick-off' of a football match from the perimeter of a large stadium will be able to draw upon that knowledge to see how the nucleus is very small in comparison to an atom.

The textbook example about comparing atoms and bricks demonstrates how the comparison needs to be explicit. The example of the atom-as-a-tiny-solar-system (see Chapter 7) reminds us that students are not always as knowledgeable about the analogue as teachers may expect. Once again, the lesson here is not to assume, but to check, that students do appreciate the aspects of the analogue being referred to. Not only do analogies have to be explicit, they also have to be selected so that the analogue is genuinely familiar to the students.

Selecting a suitable and familiar analogue is only part of the process of making the unfamiliar familiar. If students are to benefit from the comparison they should be asked to focus on both the aspects of the analogy where the analogue maps onto the target, and those where it does not. Perhaps in the Wembley Stadium example it would be clear that only the idea of relative scale is relevant. However, in the comparison between the atom and the solar system there are a number of significant differences, such as the way electrons (unlike planets) repel each other, or the way the solar system (unlike an atom) is almost planar.

This can be well illustrated by an example of a teaching analogy in use. I was observing a lesson being taught by a student teacher. He was an intelligent, enthusiastic trainee, and had prepared his lesson well. He was teaching a class of 14–15 year olds about enzymes. Large molecules to be digested had been modelled as a string of beads, which could not pass through the gut wall until the string had been broken. At one point an analogy was made between the enzyme and a pair of scissors. At this point I made a comment in my observation notes:

'In what ways is/isn't the lipase like a pair of scissors? A useful analogy – but how far can it be pushed?'

The analogy has the potential to be fruitful. As was pointed out in Chapter 6 many students find the (unfamiliar) molecular world difficult to understand. The notion of the enzyme having a role like scissors cutting through the string of beads is a potentially useful comparison. The students would all have experience of cutting things up with scissors, and were likely to be able to form a relevant mental image of the function of the enzyme.

However, what was lacking was a discussion of how far this analogy could usefully be taken. Clearly the mode of action of the enzyme molecule was very different to that of scissors. Although students at this level would not be expected to appreciate the detailed mechanism of enzyme action, they should appreciate that the process relies only on the shapes of the molecules involved. Scissors, of course, rely on two key physical principles – the concentration of force due to having sharp blades, and the presence of a pivot to allow leverage. In other words, a pair of scissors only works because an external agent (a) positions the scissors over the object to be cut; and (b) applies a force in the required direction. The action of an enzyme does not require any such external agent.

This particular class were set the task of producing a summary of their learning about enzymes. Perhaps unsurprisingly the idea of the molecular scissors was included in a number of the pieces of work. What was slightly more unexpected was the way that a number of the students drew diagrams showing molecule-sized scissors cutting up the long chain molecules (eg see Figure 10.6).

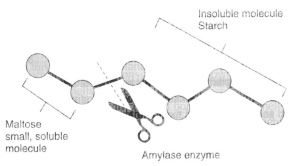

Figure 10.6 A student's diagram showing enzyme action (redrawn)

It would seem that for these students the enzyme was not a molecule which functioned like scissors so much as a molecule-sized pair of scissors. Some learning had taken place, but the task of constructing a scientific understanding was only just beginning (see Figure 10.7).

Developing a constructivist approach (summary)

Meaningful learning requires students' active participation in problem-solving and critical thinking about activities which they find relevant and engaging. They are 'constructing' their own knowledge by testing ideas and approaches based on their prior learning and experience, applying these to a new situation, and so relating the new knowledge gained to their existing conceptual frameworks.

Constructivist science teaching takes into account what we have discovered about how learning occurs. Teaching sequences should be designed that begin by eliciting the students' current ideas about a subject. This process may sometimes be carried out by the teacher's verbal questioning; or by asking students to produce concept maps, to brain-storm posters for a topic in small groups, to discuss concept cartoons, or by using a written probe as a pre-test.

Teaching can be better planned when students' alternative conceptions can be anticipated, in order to take their ideas into account. Active exploration of the limitations of students' existing thinking is often necessary to move students towards the accepted scientific view.

The design of the teaching sequence should not derive only from the content to be covered, or the standard experiments traditionally performed. Rather planning needs to be based upon an analysis of the conceptual structure that it is hoped that the students will acquire, in relation to the their current understanding. This will determine the logical sequence most appropriate for learning, and this may not match the historical development or standard textbook presentations. The students should be asked to use their existing knowledge to formulate ideas which can then be challenged by being tested for internal coherence, and for consistency with both experiment and related areas of knowledge. Clearly hypothesising, predicting, and critical discussion from an important part of this process.

The constructivist chemistry teacher is both a classroom researcher and a learning-doctor. Responding to student misconceptions effectively requires teaching that is honed to the needs of each group of learners, and is an interactive process that is both challenging and stimulating: much like science itself.

RS•C

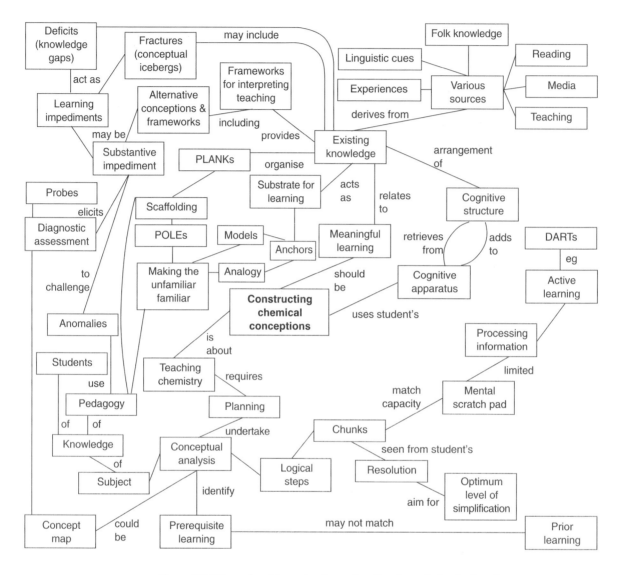

Figure 10.7 Some key ideas about constructing knowledge in the classroom

RS•C

Notes and references for Chapter 10

1. K. S. Taber, Chemistry lessons for universities?: a review of constructivist ideas, *University Chemistry Education*, 2000, 4 (2), 26–35.

2. K. S. Taber, Chlorine is an oxide, heat causes molecules to melt, and sodium reacts badly in chlorine: a survey of the background knowledge of one A level chemistry class, *School Science Review*, 1996, **78** (282), 39–48.

3. P. Black & C. Harrison, Feedback in questioning and marking: the science teacher's role in formative assessment, *School Science Review*, 2001, **82** (301), 55–61.

4. Scottish Council for Research in Education, *Taking a Closer Look: Key Ideas in Diagnostic Assessment (Revised Edition)*, Edinburgh: SCRE, 1995.

5. R. Driver, A. Squires, P. Rushworth & V. Wood-Robinson, *Making Sense of Secondary Science: Research into Children's Ideas*, London: Routledge, 1994.

6. V. Barker, Beyond appearances: students' misconceptions about basic chemical ideas: A report prepared for the Royal Society of Chemistry, London: Education Division, Royal Society of Chemistry, 2000, available at **http://www.chemsoc.org/learnnet/miscon.htm** (accessed September 2005).

7. S. Naylor & B. Keogh, *Concept Cartoons in Science Education*, Sandbach, Cheshire: Millgate House Publishers, 2000.

8. This expression should not be taken too literally. It is an open question to what extent conceptual frameworks can be said to be 'stored', or whether such ideas should be treated as purely metaphorical. One view of learning suggests that the form in which knowledge is expressed when memory traces are activated need not reflect the way knowledge is actually coded in the brain (see note 9). This may be correct, but it does not negate the usefulness of such 'metaphors of mind' (see note 10). It is certainly true that there are large variations in the extent to which different concepts are associated by an individual learner – and so the notion of conceptual structure has genuine utility, even if it does not directly reflect any underlying physiological feature.

9. W-M. Roth, Artificial neural networks for modelling, knowing and learning in science, *Journal of Research in Science Teaching*, 2000, **37** (1), 63–80.

10. R. J. Sternberg, *Metaphors of Mind: Conceptions of the Nature of Intelligence*, Cambridge: Cambridge University press, 1990.

11. K. S. Taber, The Chemical Education Research Group Lecture 2000: *Molar and molecular conceptions of research into learning chemistry: towards a synthesis*, available via Education-line, at **http://www.leeds.ac.uk/educol/** (accessed September 2005).

12. Department for Education and Employment (DfEE), Circular 4/98 *Standards for the award of qualified teacher status*, London: DfEE, 1998.

13. K. S. Taber, *Teacher – teach thyself, and then teach others by your example: some 'simple truths' about teaching and learning*, available via Education-line, at **http://www.leeds.ac.uk/educol/** (accessed September 2005).

14. A. Goodwin, Teachers' Continuing Learning of Chemistry: implications for pedagogy, *Chemistry Education: Research and Practice in Europe*, 2002, Vol. 3, No. 3, pp 345-359 available at **http://www.uoi.gr/cerp/2002_October/pdf/06Goodwin.pdf** (accessed September 2005).

15. J. Taylor, *et al*, *Science Connections 2*, London: Collins Educational, 1998, 122.

16. J. Taylor, *et al*, *Science Connections 1*, London: Collins Educational, 1997, 137.

RS•C

17. B. Milner, J. Martin & P. Evans, *Core Science 2*, Cambridge: Cambridge University Press, 1998, 107.

18. F. Harris & J. Ferguson, *Chemistry (3rd Edition)*, Harlow: Pearson Education, 1996, 176.

19. J. Taylor, *et al*, *Science Connections 2*, London: Collins Educational, 1998, 33 & 95.

20. M. Levesley, *et al*, *Exploring Science 1*, Harlow: Pearson Education, 2000, 80–89.

21. J. Taylor, *et al*, *Science Connections 1*, London: Collins Educational, 1997, 12.

22. M. Levesley, *et al*, *Exploring Science 2*, Harlow: Pearson Education Limited, 2001.

23. J. Mills & P. Evans, *Core Chemistry*, Cambridge: Cambridge University Press, 1999, 105.

24. B. Earle & L. D. R. Wilford, *Introduction to Advanced Chemistry*, London: John Murray (Publishers), 2000, 49–50.

25. K. Dobson & C. Sunley, *Co-ordinated Science 1*, London: Collins Educational, 1998, 87.

26. M. Levesley, *et al*, *Exploring Science 1*, Harlow: Pearson Education 2000, 100–101.

27. J. Mills & P. Evans, *Core Chemistry*, Cambridge: Cambridge University Press, 1999, 105.

28. K. Dobson & C. Sunley, *Co-ordinated Science 2*, London: Collins Educational, 1997, 68.

29. L. Nicholls & M. Ratcliffe, *Chemistry AS (2nd Edition)*, London: HarperCollins Publishers, 2000, 27.

30. A. H. Johnstone, Why is science difficult to learn? Things are seldom what they seem, *Journal of Computer Assisted Learning*, 1991, **7**, 75–83.

31. B. Ratcliff, *et al*, *Chemistry 1*, Cambridge: Cambridge University Press, 2000, 32.

32. J. Taylor, *et al*, *Science Connections 3*, London: Collins Educational, 1998, 109.

33. M. Jones, G. Jones & D. Acaster, *Chemistry (2nd Edition)*, Cambridge: Cambridge University Press, 1997, 31.

34. K. Dobson & C. Sunley, *Co-ordinated Science 1*, London: Collins Educational, 1998, 71.

35. J. Taylor, *et al*, *Science Connections 1*, London: Collins Educational, 1997, 90–96.

36. J. Taylor, *et al*, *Science Connections 1*, London: Collins Educational, 1997, 36.

37. P. D. Riley, *Chemistry Now! 11–14*, London: John Murray (Publishers), 1999, 179.

38. M. Jones, G. Jones & D. Acaster, *Chemistry (2nd Edition)*, Cambridge: Cambridge University Press, 1997, 121.

39. E. Lewis & M. Berry, *AS and A Level Chemistry*, Harlow: Pearson Education Limited, 2000, 49–50.

40. R. Harwood, *Chemistry*, Cambridge: Cambridge University Press, 1998, 75–76.

41. B. Ratcliff, *et al*, *Chemistry 1*, Cambridge: Cambridge University Press, 2000, 36.

42. M. Clugston & R. Flemming, *Advanced Chemistry*, Oxford: Oxford University Press, 2000, 61.

43. B. Earle & L. D. R. Wilford, *Introduction to Advanced Chemistry*, London: John Murray (Publishers), 2000, 56.

44. B. Milner & J. Martin, *Science Foundations – Chemistry (2nd Edition)*, Cambridge: Cambridge University Press, 2001, 126.

RS•C

45. N. Heslop, D. Brodie & J. Williams, *Hodder Science: Pupil's Book C*, London: Hodder & Stoughton Educational, 2000, 96.

46. B. Milner & J. Martin, *Chemistry (2nd edition)*, Cambridge: Cambridge University Press, 2001, 124.

47. N. Heslop, D. Brodie & J. Williams, *Hodder Science: Pupil's Book A*, London: Hodder & Stoughton Educational, 2000, 136.

48. J. Taylor, *et al*, *Science Connections 3*, London: Collins Educational, 1998, 109.

49. J. Taylor, *et al*, *Science Connections 3*, London: Collins Educational, 1998, 93.

50. J. Taylor, *et al*, *Science Connections 2*, London: Collins Educational, 1998, 94.

51. M. Jones, G. Jones & D. Acaster, *Chemistry (2nd Edition)*, Cambridge: Cambridge University Press, 1997, 37.

52. M. Clugston & R. Flemming, *Advanced Chemistry*, Oxford: Oxford University Press, 2000, 58.

53. K. Dobson & C. Sunley, *Co-ordinated Science 2*, London: Collins Educational, 1997, 121.

54. R. Harwood, *Chemistry*, Cambridge: Cambridge University Press, 1998, 76.

55. B. Earle & L. D. R. Wilford, *Introduction to Advanced Chemistry*, London: John Murray (Publishers), 2000, 52-53.

56. M. Clugston & R. Flemming, *Advanced Chemistry*, Oxford: Oxford University Press, 2000, 58.

57. N. Heslop, D. Brodie & J. Williams, *Hodder Science: Pupil's Book C*, London: Hodder & Stoughton Educational, 2000, 107.

58. B. Milner & J. Martin, *Science Foundations – Chemistry (2nd Edition)*, Cambridge: Cambridge University Press, 2001, 124-125.

59. J. Atkinson & C. Hibbert, *AS Chemistry for AQA*, Oxford: Heinemann Educational Publishers, 2000, 24-25.

60. E. Lewis & M. Berry, *AS and A Level Chemistry*, Harlow: Pearson Education Limited, 2000, 45.

61. N. Heslop, D. Brodie & J. Williams, *Hodder Science: Pupil's Book C*, London: Hodder & Stoughton Educational, 2000, 107.

62. B. Milner & J. Martin, *Science Foundations – Chemistry (2nd Edition)*, Cambridge: Cambridge University Press, 2001, 124-125.

63. J. Atkinson & C. Hibbert, *AS Chemistry for AQA*, Oxford: Heinemann Educational Publishers, 2000, 24-25.

64. J. Oversby, The ionic bond, *Education in Chemistry*, 1996, **33** (2), 37-38.

65. J. G. Silcock, Molecules and ions, *School Science Review*, 1961, **43** (149), 195.

66. E. Wilson, *Teaching Chemistry to KS4*, London: Hodder & Stoughton, 1999.

67. B. McDuell, *Teaching Secondary Chemistry*, London: John Murray, 2000.

68. P. Thagard, *Conceptual Revolutions*, Oxford: Princeton University Press, 1992.

69. M. Watts, Children's learning of difficult concepts in chemistry, in Atlay *et al*, *Open Chemistry*, London: Hodder & Stoughton, 213-228.

70. R. Driver & V. Oldham, A constructivist approach to curriculum development in science, *Studies in Science Education*, 1986, **13**, 105-122.

RS•C

71. M. A. Rea-Ramirez & J. Clement, Teaching for understanding part 1: Concepts of conceptual change and dissonance, *4th International Seminar on Misconceptions Research*, Santa Cruz, CA: The Meaningful Learning Research Group, 1997.

72. According to the ideas of Piaget new ideas are 'assimilated' , but may cause 'dis-equilibrium', and so lead to an 'accommodation' of existing ideas.

73. K. S. Taber, Chemistry lessons for universities?: a review of constructivist ideas, *University Chemistry Education*, 2000, **4** (2), 26–35.

74. J. Garratt, T. Overton & T. Threlfall, *A Question of Chemistry*, Harlow: Pearson Educational, 1999.

75. R. Millar, Constructive criticisms, *International Journal of Science Education*, 1989, **11** (special issue), 587–596.

76. D. Sprague & C. Dede, If I teach this way, am I doing my job? Constructivism in the classroom, 1999, available at **http://www.qeced.net/ed/Construct/Construct1.htm** (accessed September 2005).

77. K. S. Taber, Building the structural concepts of chemistry: some considerations from educational research, *Chemistry Education: Research and Practice in Europe*, 2001, **2** (2), 123–158, available at **http://www.uoi.gr/cerp/** or **http://www.rsc.org/Education/CERP/index.asp** (accessed September 2005).

78. B. Indge, M. Rowland & M. Baker, *A New Introduction to Biology*, London: Hodder & Stoughton, 2000, 45.

79. J. Boyd & W. Whitelaw, *New Understanding Science 3* (Revised National Curriculum Edition), London: John Murray (Publishers), 1997, 74.

80. K. S. Taber & M. Watts, The secret life of the chemical bond: students' anthropomorphic and animistic references to bonding, *International Journal of Science Education*, 1996, **18** (5), 557–568.

81. J. Boyd & W. Whitelaw, *New Understanding Science 3* (Revised National Curriculum Edition), London: John Murray (Publishers), 1997, 68–69.

82. A. G. Harrison & D. F. Treagust, Learning about atoms, molecules, and chemical bonds: a case study of multiple-model use in grade 11 chemistry, *Science Education*, 2000, **84**, 352–381.

83. This statement appeared in the homework of one of the students in the class of 14–15 year olds discussed in this section.

RS•C

Keywords index

RS•C

General reading

For an introduction to learners' ideas in the 11-16 science curriculum

R. Driver, A. Squires, P. Rushworth & V. Wood-Robinson, *Making Sense of Secondary Science: research into children's ideas*, London: Routledge, 1994.

For a large set of Concept Cartoons suitable for eliciting alternative conceptions and encouraging discussion of scientific topics for students (particularly those up to age 14):

S. Naylor & B. Keogh, *Concept Cartoons in Science Education*, Sandbach, Cheshire: Millgate House Publishers, 2000.

For useful books about teaching chemistry 'constructively':

J. D. Heron, *The Chemistry Classroom: Formulae for Successful Teaching*, Washington, DC: American Chemical Society, 1996.

K. Ross, L. Lakin & P. Callaghan, *Teaching Secondary Science: Constructing Meaning and Developing Understanding*, London: David Fulton Publishers, 2000.

J. J. Mintzes, J. H, Wandersee & J. D. Novak, *Teaching Science for Understanding: A Human Constructivist View*, London: Academic Press, 1998.